Also by Philip J. Hilts

Scientific Temperaments

Behavior Mod

Memory's Ghost

The Nature of Memory
and the Strange Tale of Mr. M.

Philip J. Hilts

A Touchstone Book
Published by Simon & Schuster
New York London Toronto Sydney Tokyo Singapore

TOUCHSTONE
Rockefeller Center
1230 Avenue of the Americas
New York, NY 10020

First Touchstone Edition 1996

TOUCHSTONE *and colophon are registered*
trademarks of Simon & Schuster Inc.

Designed by Levavi & Levavi

Manufactured in the United States of America

1 3 5 7 9 10 8 6 4 2

Library of Congress Cataloging-in-Publication Data
Hilts, Philip J.
Memory's ghost : the nature of memory and the strange tale of Mr. M. /
Philip J. Hilts.—1st Touchstone ed.
p. cm.
"A Touchstone book."
Includes bibliographical references and index.
1. Memory disorders—Case studies. 2. Memory.
3. Frontal lobotomy—Complications—Case studies. I. Title.
[RC394.M46H53 1996]
616.8'4—dc20   96-13218
ISBN 0-684-80356-9
0-684-82356-X (Pbk)

To H.M.:

All the conduits of my blood froze up,
Yet, hath my night of life some memory,
My wasting lamps some fading glimmer left,
My dull deaf ears a little use to hear:
All these old witnesses, I cannot err,
Tell me thou art my son . . .

—SHAKESPEARE,
Comedy of Errors

Preface

Memory, so familiar to us as a private act, becomes strange as a subject of inquiry. We know it personally, as we know our limbs; we depend on it as we depend on food or shelter; we enjoy it as we enjoy a comfortable pair of shoes and favorite suit of clothes.

But we can hardly imagine what it is. We have trouble linking our own experience of memory with the scientific-sounding explanations involving computer storage, brain parts and cellular chemistry.

Now there is a greater reason to try to find the way to explain these events of experience. Just in the past two decades, the ways of memory have begun to be understood for the first time in detail and greater progress has been made in that time than in the previous three thousand.

The first efforts to relate this news have, for the most part, failed to go beyond technical discussions. They have related little of the history of memory, they have not explained why humans have the skill in the first place, and they have not described it working in natural human settings. I think to understand the subject in its new light, we need to glimpse the literature and history of the subject, in addition to the science. So, I have here tried to view the science of memory with a writer's eye.

Because of these considerations, and practical ones, I have chosen an unconventional mix of literary style and scientific material in this book, to try to give life to a subject I find compelling. Ultimately, we must obtain understanding from both knowing and feeling, from science and literature, and so I have tried here to mix those volatile elements.

The book which follows has its hazards; it enters the territory of two warring tribes, those literary and those scientific. Scientists may be displeased at points left vague and details sacrificed to keep the narrative afoot. Those readers of literary bent may want more of the life of Mr. M. that is just not available. (The book might well be called a biography of a man with no life.) Others will want a closer examination of the scientists here: what did they say here, and what was their motive there.

The subject of the book is memory, described as an exploration of one case and the side-journeys that case prompted. As there are already books describing the scientific detail, more of that is not essential here. As for the literary values, more personal detail about H.M. would certainly help make a fuller portrait, but it simply cannot be obtained within practical limits.

As for the question of how this subject relates to modern psychiatry, I think that the rapidly expanding field of neuroscience will gradually create a new psychology. This does not mean that Dr. Freud, who championed inner mental life, or Dr. Watson, who concentrated on behavior, must be left behind. Rather, we must incorporate the insights of past thinking into new forms. We realize now that Dr. Freud was incorrect in his assessment of memory: it is not a complete record waiting to be unearthed. On the other hand, his humane approach which put each person at the center of his own struggle, and set the professional therapist as a guide, need not depend on such a mistaken detail. The human-centered approach of Dr. Freud was the beginning, not the end, of psychiatry.

(In therapy, we now know, we are in fact creating the meaning of our memories, recasting them in new roles, so that we may understand the story of our lives in a positive manner, a manner that permits us to get on with living. For patients, the recreated meaning of our memories may be more important than the memories themselves. Of course, outside the personal therapy session, these memories may be dangerously misleading if taken as fact. They can be used reliably only privately. We make the stories of our lives from both fact and imagination.)

The person at the center of this book is Mr. M. himself, and what appears here includes the memories of Mr. M. as he tells them. I have chosen to believe these reminiscences in the narrative, though he is inconsistent in his recollections. I use the most prominent versions he describes, checked where possible against known facts. There is so little of his life we have, and it is *all* he has, that I felt that the only reasonable and dignified path was to accept some of what he remembers, uncertain though it may be.

So follows an account from science and literature, with details personal and impersonal detail, with facts certain and uncertain, in an effort to tell the tale of memory and Mr. M.

Part I

1.

I recall standing on the bank of the Charles River, brooding
on memory. The day was sunny, and I peered into the dark green
waters where there were pale stones on the bottom, lit as if by
green moonlight. These are like memories, faces beneath the
surface, now clear, now blurred, looking up at us, mouthing
words we may or may not understand.

I wanted an explanation of this phenomenon, of just how the
meat of the brain can give rise to these spirits of memories. I
was unable at first to picture it, this insubstantial substance, this
unbodied body of thought. It seemed impossible in some ways.
I have many times dropped the subject, only to pick it up again.
Like a stray dog, the subject of memory follows me home. I
don't know why it's there, and I can't shake it.

So I have made a habit of asking about memory when I meet
someone apparently knowledgeable, or when I read apparently
authoritative text. But finally, the explanations left me unsatisfied. I read Marcel Proust and was charmed by his prose as he
spoke of a memory: "When from a long distant past nothing
subsists, after the people are dead, after the things are broken
and scattered; still, alone, more fragile, but with more vitality,

more persistent, more faithful, the smell and taste of things remain poised a long time, like souls, ready to remind us, waiting and hoping for their moment." So his lovely meanderings go, but without any explanation of *how* they go. The texture of memory we can feel easily enough, but a sensible explanation of it is more difficult.

One of the moments I resolved to pursue the question was at the bedside of my wife, who was dying needlessly, mother of small children. We don't always feel we need to remember— young people seem to disdain the impulse—but sometimes the need arises forcibly. I felt that imperative as I sat on a stool beside the bed, when she spoke and gestured that she finally had lived enough, and wanted an end. Memory, above all other things, is personal; sometimes it feels as mysterious and purpose-filled as sex and birth. I sat in the hospital room, where poetry never intrudes, under the stare of busy doctors, and wrote a few lines of poetry, the memorious impulse overtaking me.

Again later I resolved to inquire more deeply into memory. One afternoon, in the brown-and-gray time between light and night, I read a passage which bolted my attention to the page.

I was alone in my study reading Samuel Johnson, that melancholy and tic-infested man of eighteenth-century letters. Dr. Johnson had a bursting intellect bound in a crippled body. It was partly because of this that he cherished his memory and understood it better than anyone in his age, for he spent a great part of his waking hours within it.

Johnson noticed something different between the behavior of humans and that of animals. Animals—and here he tactfully did not include other men, but only animals—seem always to be fully employed mentally, "or to be completely at ease without employment . . . they feel few intellectual miseries or pleasures." They have no exuberance of understanding, but have their minds exactly adapted to their bodies.

But humans do not have their minds and bodies so efficiently matched. "So few of the hours of life are filled up with objects adequate to the mind of man," he mused, "and so frequently do we lack present pleasure or employment, we are forced to have recourse every moment to the past and future for supplemental satisfactions."

Humans have more candlepower than we feel comfortable with. It is this excess of power, he said, this surfeit of sense, which is precisely the power of memory. It is the ground from which human culture arises. "It is indeed the faculty of remembrance which may be said to place [humans] in the class of moral agents," Johnson wrote. Reaching the full pitch of enthusiasm in his insight, he added, "Almost all we can be said to enjoy is past or future," and springs, therefore, from memory. It is the originator of both goodness and of grief.

Repeating a thought from his predecessor, John Dryden, who in turn pulled it from Horace, he said that memory is the one district of life, the one province, specially beyond the reach of both man and God: "Fair or foul . . . the joys I have possess'd in spite of fate are mine./ Not heaven itself upon the past has pow'r / But what has been has been, and I have had my hour."

It is not always welcome, he knew. That "snake memory," Shelley spat, when he felt abused by it. It is dangerous, said Proust, "a kind of pharmacy or chemical laboratory in which chance guides our hand now to a calming drug, and now to a dangerous poison." The poet Walter de la Mare described memory as a woman he disliked. "The strange deceiver. Who can trust her? How believe her?—while she hoards with equal care the poor and trivial, rich and rare; yet flings away, as wantonly, grave fact and loveliest fantasy? . . . So bewitched her amber is, it will keep enshrined the tiniest flies—instants of childhood, fresh as when my virgin sense perceived them—And can, with probe as keen, restore some fear, or woe, when I was four."

The point which most intrigued me was that Johnson realized that memory not only re-creates the past, but must be the source of the future as well, because all our salients into the future are built upon what we know from the past, and what we fancy from it. Conferring with memory's ghosts, consulting its tables of facts, we project the future and what we expect it to look like. Memory makes us, fore and aft.

The bright edge of consciousness moves quickly, Johnson said, and the present, after all, is in perpetual motion, a precarious ledge: "It leaves us as soon as it arrives, ceases to be present before its presence is well perceived, and is only known to have existed at all by the effects which it leaves behind. The greatest part of our ideas arise, therefore, from the view before us or behind us."

This struck me as profound. I began to imagine myself in the act of thinking, imagining a river of time, and trying to cut from it just the present moment. The present is just that scoopful of time I can attend to at one glance, before the head of it proceeds into a fog. William James, the colossus of American psychology, actually tried to quantify the "present" during a little reflective experiment at the end of the last century. He said that the present may be said to be about three to twelve seconds long. Seconds! All else is memory, and reconstruction from it.

I would have gone no farther on the subject of memory probably, but for one afternoon when I heard of one particular case, the case I think must be the most remarkable in the annals of human memory. From then, I could not help but throw myself into the subject as a full avocation in my spare hours.

It was a summer afternoon before a thunderstorm when, sitting alone at my desk in Washington, D.C., among other empty desks in the newsroom nearby, I first heard of H.M., as he is called in the scientific literature. The small hairs on the nape of my neck stood up. I became still with surprise as, on the other end of the telephone, a voice described the rumor of his existence and in a matter-of-fact tone told of his condition: There was a man she had heard of, a patient in the hands of curious doctors, who had had his memory removed. This of course seemed impossible at first. I was told that he could still think and converse, as long as those actions were held within the small circle of the present, a few minutes. But once the present passed on, he could hold on to nothing. He forgot with whom he was speaking, and what the subject was. Nor, for that matter, did he know where he was. For the most part, he could remember no one he has met, nor anything that has happened to him.

What a curious loss! If memory, which makes up the very bones of thought, could be so isolated, and so selectively amputated from a man's mind, what was there to give his mind any human shape at all? Would he not collapse back into an animal's life, all sensation and reaction, without any savor of reflection, without the favor of any planning? There would be for him nothing upon which he could build a past, or a future. And, still more disturbing, he would be unable to form any human relationships.

To my friend on the telephone, the case did not seem that extraordinary; she was merely relaying a bit of information, a curious case, and back to work. I hung up, rose from my desk, and stood for a moment looking past heaps of papers through the large windowpane across a small waiting area, out into the city. I looked at the building across the way—in the midst of construction, destruction, or renovation—after the crane and ball had hollowed out the interior completely, while leaving the shell to stand alone. From a certain perspective in the street, the facade was intact, the building appeared solid, a place one could enter, rise through the floors, and conduct business in the layers and cubes familiar to urban people. But from another view, my view, four floors up and across an alley, the inside was empty, an archaeological marvel revealing the secrets of its fabrication: organs of metal and wire which had been snug between floors lay exposed; sinews which had run up and down between the walls were now stripped and dangling. (Don't we see one another this way? I see you solid as a building, a full and seamless character with whom I exchange business and love, without looking for the tubes, beams, and wires that underlie the mechinics of your behavior. Though we have spent centuries pondering the questions, I still cannot tap on your walls and discover by the hollow or firm sounds which of your walls are merely decorative, and which ones hold everything up.)

In the person of this man, Mr. M., with the interior of his humanness so neatly deleted, I thought I glimpsed a chance to see something of what lies behind the screen of appearances. Here was, in a paraphrase of Anatole Broyard, an opportunity to examine, at point-blank range, the threatened human mind and soul. Who was he? Where was he? Could it be true?

Memories are hunting-horns
whose noise dies away in the wind.

—Guillaume Apollinaire,
"Cors de Chasse"

2.

I began by laying aside what information I could acquire in bits, while I continued with my daily work and family life. The confirmation of what I had first heard did not come immediately.

I had to wait until one fall afternoon, when I visited the wooded campus of the National Institutes of Health in Maryland. There, a freckled and somewhat shy woman, Dr. Suzanne Corkin of M.I.T., was giving a short talk on her studies in memory. Few people were present in the small classroom when she spoke of the current condition of that unusual patient, Mr. M., whom, I found, she herself had studied. Moreover, she was his scientific overseer and guardian. She spoke a little about brain structures and varieties of memory in this patient; the language seemed opaque to me at the time. But she added some human details. Because of his condition, she said, it was possible for him to read a newspaper or magazine, put it down, and then twenty minutes later pick it up fresh again. He might well read it over and over, each time for the first time. The date on the cover would be no help, as he hadn't the faintest notion of what the day, month, or year was. I asked a few questions. Each morning, she said, he woke and did not know where he was. He could not recall who was taking care of him, though the nurses had attended him for many years steadily. He had been in this state since 1953, and did not know that fact either. There was almost nothing before or behind him; he was perpetually held in one moment, a moment resembling that instant when we wake from a particularly heavy sleep and cannot recall just where we are, or just what business we are about.

After the talk, I spoke to Dr. Corkin and told her I wanted to find out more about him. "If you are thinking you want to talk to him, forget it. We protect him very carefully," she said. The answer was discouraging. But I did have something to start with; I could read the scientific papers which were written about him, though they are in the language of the scientific journals, which seeks to make the narrowest possible footprints as it moves, perhaps for fear of being caught in some misstatement. There were virtually no human details of his life recorded, nor any full description of what had happened to him, but only accounts of the tests done to explore the cavity that had been his memory.

Perhaps I should say that I also knew by this time that his loss was not an accident, but had occurred from a deliberate and poorly conceived clinical experiment. This thought crosses my mind as the shadow of a cloud. I drove home from the campus of the Institutes, feeling I had never known of a human situation which seemed so curious. Questions crowded in, stumbling over one another. Would he feel lost, or angry? Would he be frightened, awakening every day not knowing where he was or what had happened? What substance was there in his life, if all was water and sand running through his fingers?

I read on, and found that after the case of H.M. had occurred and been passed around within scientific circles, in several universities there had been a sudden rush of interest in the subject of memory. Other advances in the study of the brain came simultaneously, and there had even been a revival in a traditionally soft and sleepy subject, dormant since the nineteenth century, the study of human consciousness. In notes and references in a number of books and articles, I found that the renaissance in memory science could be said to start from Mr. M., as it was his deficit which gave some first crucial clues, and it was work on what happened to him that was the first of the coming rush of research. He has become, over the past forty years, a famous patient in psychology, a *cas célèbre*, and is counted as the patient most studied in the history of medical literature.

At the moment in midcentury when he appeared, science was prepared for quick progress, and his case was the pivotal moment when the study of memory—carried on so long in the realms of the philosophical and metaphysical—finally turned toward the biological. At the time, psychology was still dominated by the ideas of men who were essentially adventurers of the previous century: Sigmund Freud, who suggested there must be much beyond conscious experience; and John Broadus Watson, who viewed matters from the opposite perspective, and suggested that simple contingencies in the world shape our behavior regardless of any internal mental life.

But theoretical notions about the mind remained shaky and, as a signal that they were not mature, they were essentially useless in the treatment of severe mental disorders. (A revolution in the treatment of mental illness was indeed coming, just at midcentury, but not as a result of flights of theory; rather, it came from a few clinical accidents. By both luck and experimentation, doctors stumbled on five drugs to treat schizophrenia and depression. They were practical and effective, though still limited. One was reserpine, drawn from the plant *Rauwolfia serpentina*, which the ancient Hindus had used to treat both insomnia and insanity. Indian physicians brought the drug into the modern pharmacopoeia when they discovered that it reduced high blood pressure; an American psychiatrist rediscovered what the Hindus had suspected, that it could calm schizophrenics. Another drug, chlorpromazine—first synthesized as an antihistamine, a treatment for hay fever or allergy—also seemed effective with schizophrenic patients. Iproniazid, an antibacterial agent, sometimes improved the mood of people taking it for tuberculosis, and thus an antidepressant was created.)

In the field of memory, the most fashionable notion then gaining currency was that of Dr. Karl Lashley. It was unknown what parts of the brain were used for what thoughts, and in the 1930s Lashley had begun a quest of many years. He reasoned that memory must be located somewhere. So, if parts were removed from rats' and monkeys' brains systematically, bit by bit, eventually he would strike the part that contained memory. When it was removed, the animals' memories would fail. Through thousands of experiments he pursued the engram, the brain's actual record of memory. To obliterate it would be to find it, but he could find nothing short of total destruction of the brain that would prevent rats and monkeys from remembering. He concluded with frustration in 1950 that as far as he could tell memory itself was not possible: "This series of experiments has yielded a good bit of information about what and where the memory trace is not. I sometimes feel . . . that the necessary conclusion is that learning is just not possible."

But the peculiarities of chance and history enter here, and from stage left comes Mr. M.

Experiments of brain removal did not work in rats and monkeys, but in one human experiment the strategy succeeded by chance. On a summer morning in 1953, in Hartford, Connecticut, Mr. M. went into surgery and central portions of his brain were sucked out by a silver straw. The surgeon hoped that the areas removed might be the source of Mr. M.'s worsening epilepsy, and excising them might relieve it. Unfortunately, the epilepsy did not disappear, though it did improve significantly. But immediately after Mr. M. left the recovery room it was clear that something else was dreadfully wrong. He was uncomprehending. Yes, he could speak and read and converse. But when asked where he was, and who were these people at his bedside, he did not know. He could not find his way to the bathroom, and nurses could enter the room, speak to him, leave, and then return a moment later to find he had no memory of them. This was a demonstration, in the flesh, that there is at least one actual organ of memory in the brain, and that it could be named and circumscribed.

Scientists, beginning with Dr. Brenda Milner of Canada, an expert on memory, and the surgeon who caused the catastrophe in the first place, Dr. William Scoville, soon began a search to understand what had happened to Mr. M. and what it might mean. Dr. Scoville and others went back to excising parts of animal brains in hopes of duplicating the striking damage found in Mr. M. Psychological tests were done on Mr. M. which have continued to this day, and which have attempted to outline what abilities he has left and what he lost on that morning.

Over decades, one thing led to another, and this work joined the streams of other work in progress to become a river of new research. There was a rapid elucidation of the biochemical workings of that part of the brain which had been damaged in Mr. M. Cell by cell, maps of brain areas were made. Now, some decades later, we have seen the first image of a human memory being formed, as it was captured by a scanner. In recent months, in fact, the tale has circled fully round. Scientists in California and New York have marshaled the information obtained since Mr. M.'s case began, and have created a drug which acts on the cells within the memory organ destroyed in Mr. M. The drug has succeeded, in a variety of animals so far, in enhancing for the first time the natural memory system of the brain.

(The case brought strongly to mind images of another man with a brain injury who underwent extensive questioning and analysis. Phineas P. Gage, in the fall of 1848, was a railroad worker of some intelligence and good humor, who, when distracted, used a three-and-one-half-foot steel tamping rod to tamp down explosive powder within a drill hole. He had not noticed that the sand usually used to buffer his blows with the rod upon the powder had not been poured in the drill hole. He thrust the rod down hard upon the powder; the steel rod blew out of the hole, a missile which penetrated his skull below the cheekbone and then fired out through the top of his head. The accident left him entirely intact intellectually and physically. He had no memory loss, and his ability to learn was still strong. But he lost control of his emotions; a mild man before, he became blustery and full of curses. He had been honest and kept his word; now he was unable to be counted on. He began a boisterous, wandering life. The study of him suggested much about the front parts of the brain, where his brain was disrupted. It appeared to be the seat of rational control of behavior. Drs. Hanna and Antonio Damasio, in a recent resurrection of the evidence in that case, wrote that what had been disturbed in Mr. Gage was his moral reasoning and social control, which thus must be located in the front center of the brain. The conclusion had been unacceptable in the last century for various reasons. I thought that in Mr. M., whose brain damage was also complete, and also narrow and circumscribed, there had appeared another case in which substantial insight was possible.)

This story has the flavor of both scientific and literary concern, an attempt to present both facts and feeling. On the side of human feeling and experience, I find it extraordinary that in mythology, this moment was prefigured. This particular loss was foreseen.

In the Greek, Mr. M. would have been plunged into the cold waters of Lethe, the river of forgetfulness. To the ancient Greeks Memory was not a concept, but more, a goddess. Her name is Mnemosyne, her power was over mind and imagination, and she became mother of the nine Muses whom she conceived in nine nights with Zeus. She reminded men of their dead, their heroes, their place in the world; she presided over the arts of expression, and poets thus became the "masters of truth." It is memory that is the antidote for oblivion: Mnemosyne knows "all that has been, all that is, all that will be," says Hesiod. Remembrance in the Greek tales was not a tool to set events in temporal frames for analysis as we see it, says Mircea Eliade, but was used to reach deep in the black mud of being, to discover the "primordial reality from which the cosmos issued."

In India, no less so. There, our lost man would be the yogi who, while traveling, fell in love with a queen. Enthralled by her beauty, he went with her to her country and lived in her palace, completely forgetting his identity as a man of purity and understanding. On a trip to the spirit world a student heard of his master's enslavement, and realized that his master was doomed to die because of this forgetfulness. In the realm of death, the student searched the Book of Fates, found the leaf containing the destiny of his guru, and erased his name from the list of the dead. He then raced to Ceylon, and disguised as a dancing girl, sang sweetly to his master. Little by little, the master remembered his true identity, understood that his oblivion was, in its essence, forgetfulness of his true and immortal nature. Thus, writes Eliade, Indian literature uses images of binding, chaining, and captivity interchangeably with those of forgetting. When the gods lose their memories, they will fall from heaven.

The Celtic tale is more earthy, and more like the actual condition of Mr. M. In the Irish, M. is Sweeney astray, the warrior exiled from the rest of humanity and turned to a bird. He had lost all, as Seamus Heaney's translation goes, "My people . . . kith and kin, where did their affection go? In my heyday, on horseback, I rode high into my own: now memory's an unbroken horse that rears and suddenly throws me down. Over the moors and starlit plains . . . to his cold and lonely station, the shadow of Sweeney goes."

Thus was Mr. M. veiled with forgetfulness of his own nature, and became lost. It is memory that raises man up from the animal life, and from that summer day, he has walked among us only as a shade, a man banished from the company of the species, sent out from the tribe of the memorious. He is with us now only as the object of study.

In literature, Mr. M. might take his place beside Joseph K. in Kafka's *The Trial*, or in the gallery of lost souls that Samuel Beckett has brought to life. Mr. M., as Jerome Bruner says of another patient, is a character "dispossessed of the power to find meaning in the world. And, rather than dismissing him as beyond the pale of human explication, we must ask instead about his subjective landscape, his implicit epistemology, his presuppositions."

In the case of Mr. M. we have a chance to observe, in both human and intellectual terms, the story of an individual tragedy and of how knowledge lurched forward from that tragedy.

Over a decade, I pieced together evidence about what happened to Mr. M. and what happened because of him. Dr. Suzanne Corkin, the figure most central to the life of Mr. M. over the past twenty-five years, at first declined to help, fearing an avalanche of interest in him which would bury his current peace and solitude. But somehow over time, she came to trust what I wanted to do, that is, to write a simple, sober account of what had happened to him and what it may mean. We finally spoke at some length and she began to open his records to me.

After six years of reading scientific papers and visiting laboratories, I stiffed my back and went to see Dr. Corkin again. I wanted to talk more; more, I thought, than she could possibly agree to. But she was willing, and even suggested that it would not be out of bounds for me to meet Mr. M., if I could get the approval of Dr. Brenda Milner, who was the first scientist to study him and was in some sense his guardian, like a grandmother of the spirit. She agreed readily, more readily than I had expected, and I began a number of visits to M.I.T., and the home where Mr. M. now spends most of his days, to talk with him. More of that later.

The chief characteristic of the mind is
to be constantly describing itself.

—Henri Focillon

3. &

I visited the Marine Biological Laboratory in Woods Hole, Massachusetts, one June on a fellowship. I recall being sent to fetch sea snails from the wooden hut where sea species of all kinds were kept in tanks. It was a sharp day on the shoulder of Cape Cod, and a chilly breeze, though it was summer, cut through my shirt as I walked toward the door, then went up the creaky stairs, to the second floor of the dark-halled building, down a corridor, and into a small storeroom. There in rows on the metal shelves were clear plastic boxes two feet long, filled with cold seawater. I slid one box from its shelf and peered into the fluid. A dozen small animals, each translucent and about as long as my thumb, were clinging to the bottom of the little tank. As I lifted down the box, light from the window struck the backs of the animals and from them came flashes of orange and blue. These were the subjects of memory studies, sea snails, nested near the other animals with simple memories, the squid and the jellyfish.

Memory is one of those words familiar to all of us, but which becomes confused when technical knowledge begins to accumulate around it. I thought I should understand its sense in both literary and scientific language or I would never be sure of either. We can't speak of memory in any exact terms and still be plain, I suppose, as we are still so unfamiliar with the gray medium in which it lives. We resort to analogies—the computer and its electronic registers, the enchanted loom, and the lights moving beneath the surface of a pool.

In terms that fit both art and science (and there is little that fits both), memory is said to be an effect, more or less permanent, that remains after a presence to the senses has gone; it sounds like the description of a ghost. I began some of my adventures in the technical realms of memory, with small sea animals. They can give an idea of what the simplest memory systems, or pre-memory systems, might be like, and they also suggest some answers to the historical question of what memory is for in the first place.

Watching the simple animals is like looking through a telescope back into time, into the early moments of our beginning. The neurons of these animals are large and few in number—more primitive but easy to work with at the bench. Fortunately, they are sufficiently like the comparable filaments in mammals, so we can make more or less immediate translations of discovery from lower to higher species. The Woods Hole labs are especially famous for squid, which have gigantic neurons, visible to the unaided eye and so broad that electric wires, even in the days before such wires were refined to a micron across, could be placed simultaneously inside and outside the cell, to record its electrical state and the waves of charge passing through it.

In the hallways at the marine laboratory, you can sometimes hear something clicking rapidly, giving off a sound like an old prop plane, as the activity of live neurons is broadcast by a speaker. These bursts are the sound of neurons probing and reacting to the world, and call to mind the echoes of bats as they fly, or the clicks and whirs of dolphins sketching acoustically the landscapes that surround them.

It was in the labs and in the century-old library that I began to see the history of memory as starting in primordial tidepools.

Before the first bits of living flesh, the earth had no perceiving creatures, and of course no memories. The planet had never been experienced by living things. To our later minds this was simply a world "without form and void" out of which a firmament was later created. The firmament of Genesis, I think, was not so much solid land as the first inner imagining of land. It refers to the first internal "image" of the world, which living things have carried since.

The first animals behaved, but they could not form even the simplest association between, say, dark shapes and danger. There are still with us creatures like this, which do not use neurons and thus cannot react on the moment or with any sophistication. A sponge does not behave, says neuroscientist Dr. Richard Thompson, it just rests on the bottom of the sea. If nutrients are in the water, it lives; if not, it dies. But for the jellyfish, behavior became possible; it has strung nets of signaling cells across its clear globe. Information from one part of its skin is relayed to other parts, and causes reactions there. Thus experience builds: light pierces a membrane. Waves of sound pat against it. Solid forms press on it. Some of these events were, generally speaking, a good thing, others were neutral, some were bad. As Nicholas Humphrey explains: An animal that had the means to sort out the good from the bad—approaching or letting in the good, avoiding or blocking the bad—would clearly have been at a biological advantage. Natural selection was therefore likely to favor sensitivity. Being sensitive need have meant, to begin with, nothing more complicated than being locally reactive, in other words, responding selectively at the place where the surface stimulus occurred. Just as today we might say that a person is sensitive to the sun if he responds to sunlight on his neck with local reddening, so the first types of sensitivity would have involved, for example, local retraction or swelling or engulfing by the skin.

The next step toward memory was the ability not to respond immediately but to wait briefly after a stimulus, and call up former associations before determining what to do next. External impressions are laid down internally, leading to predispositions, tendencies to feel or react in a certain way. In the wet interior of the mind, the impressions accrete and in these deposits we can sense pre-experience, the pre-thought of minute animals.

Their reach eventually extended beyond their grasp as they sensed things at a distance—instead of sensing their prey or their hunters directly, they began to detect them by their chemical signatures upon the water, or by noting the varying illumination they scattered off their bodies, or perhaps by sensing the drumming, whistling vibrations carried in the currents.

Memory is therefore a system for guiding reaction; the twitches of muscles and the squirts of glands can be far more effective if they are timed and shaped to deal with passing threats and opportunities. This is, as Herbert Spencer wrote, the basis not only of memory but of all psychology, the great principle of mental life—"the adjustment of inner to outer relations."

The working unit of that first simple association was formed a billion years ago, when some cells of the ancient jellyfish began to specialize, to become neurons. It is a pretty fact that the neurons themselves—signaling cells—remain rather simple and have changed little in the intervening several billion years since they first specialized. They are also little different from species to species among those which now use them. The creation of true thought and elaborate behavior comes not from the sophistication of the basic working parts but rather from larger and larger numbers, joining in more and more elaborate patterns to create new effects.

Sea slugs and squids as models of the simplest memory began to fascinate science in the 1960s and 1970s, shortly after the accidents which befell Mr. M. I visited two laboratories where work on these species has been carried on over the past few decades—the Marine Biological Laboratory in Woods Hole and Columbia University in New York City. The shell-less snails I watched by the window at the Marine Biological Laboratory were *Hermissenda crassicornis*, a color-saturated beauty from cold Pacific waters brought to the marine labs for the inquiries of Dr. Daniel Alkon. This snail is, incidentally, among the most beautiful in anyone's list of species on land or sea. Its skin is often translucent, with billows in skirts like Spanish dancers, or streamers of soft filaments upon its back. The folds and ribbons of skin are brilliantly hued, with tones unlike anything seen outside the waters of the tropical fish—vibrating red, intense yellow, pearlescent green, black-light purple, electric blue, and neon orange. Its terrestrial equivalent is the butterfly, but there are more types of these sea animals, each more surprisingly marked with stripes, dots, regular and irregular curves. The sea slug, more graceful and active than its cousin the garden slug, glides more quickly and turns its head as it moves like a dog sniffing out a path. It lives for the most part on sea grasses and soft corals near the water's surface, sliding upward on slim limbs toward light and food, or fleeing downward into the dark water away from danger.

It has few neurons to rub together, but is an absorbing little mentality. The creature has only five neurons in each eye, backed up by thirteen in each of its tiny optical ganglia. Bigger snails may have five thousand such cells, and rabbits have nine hundred thousand. Human visual systems, hundreds of millions. Nevertheless, it took Dr. Alkon thousands of experiments and most of his life to map *Hermissenda*'s eye neurons and all their connections—in effect, to create the complete wiring diagram for the *Hermissenda*'s eyes.

As Dr. Alkon writes of his years observing it, a chance encounter by two *Hermissenda* often "began with a succession of rapid, brief touches, which allowed each animal's pair of finger-like tentacles to assess the size, and perhaps the intentions, of its potential opponent." If one of the animals were much smaller, it would turn and retreat, while the larger "might rear up and pounce, as a cat might lunge to engulf a mouse. Two animals of comparable size might engage in a series of brisk rearing and lunging movements strikingly reminiscent of two stags competing for the attention of a nearby doe. Once one was clearly at a disadvantage in such a contest, it executed a violent writhing movement, thrusting itself off the aquarial wall and hurtling somewhat unpredictably toward the coral gravel below. However elementary its nervous system, this animal was capable of impressively elaborate behavior."

Clinging to the plastic sides of its laboratory harbor, *Hermissenda* seems to me a fragment of the wiring needed for memory, one of evolution's early impulses leading to thought as similar simple engines joined to make more elaborate creatures. Thus these creatures are part of our history, the history of memory.

(Perhaps a similar, though artificial, evolution is what we are now seeing in computing. Though literature has imagined robots in human form, like the haunting simpletons of Fritz Lang's *Metropolis*, these will never materialize. Instead, machine intelligence, in the form of little insectlike devices spread throughout our daily lives, are as close to robots as we will get. Evolving in human hands, these become distributed bits of intelligence such as thermostats and timers, telephones with little minds to tell us who is calling, and send calls forward to the phone nearest us. Seen this way, it will be a long time of evolutionary accumulation before we may know whether artificial intelligences can reach the level of animal consciousness or higher, self-reflective consciousness.)

Shortly after I spent the summer in Woods Hole being introduced to the tiny and elegant *Hermissenda*, I paid a visit to the more famous haven of sea snails at Columbia, at the laboratories of Dr. Eric Kandel. These were not seaside, as Woods Hole labs are, but cliff-dwelling, above a boulevard. In Kandel's lab the sea snails, though related, are so large by the standards of the species that they are called sea hares—softball-sized and anything but beautiful. The sea hare is dark, spotted, a brown-and-purple color to its slimy body; its name comes from its large protrusions that look like fleshy antennae, or small rabbit's ears, at its fore end; around its body is a ruffled skirt of flesh. It is this creature, *Aplysia californica* in Latin, which Kandel was introduced to by a French scientist, Ladislav Tauc, and with which Kandel in 1963 began the small branch of neuroscience in which the roots of memory were to be sought by examining its simplest examples in the simplest creatures. He also reached back to the tidepool for inspiration. He found that when he touched the ruffled mantle on the sea hare's body, it and the animal's fleshy gills would quickly retract in a defensive maneuver.

Kandel and Tauc first thought of sounding a tone, or flashing a light, so that a classical Pavlovian condition might be set up. Eventually, just the tone or light would induce the withdrawal. But they failed at this, and found another little piece of memory to work with—touching the mantle once got a strong withdrawal reaction, but if the mantle was touched ten times in a row, eventually the animal got used to the touch and didn't bother to pull its mantle all the way in. This is habituation, as author Susan Allport noted, and it is an extremely useful phenomenon because it allows us to ignore stimuli that have lost novelty or meaning—the clothes on our backs, traffic noise, the sensation of grass between our toes—and get on with the business of living, without constant alarms of sensation.

The idea was that Kandel and his group would map every neuron involved in the sensing-and-withdrawal reaction. Then they would try to find what inside the neuron changed with the learning. They squirted jets of water at the sea hare (to make the squirts uniform, they began using a Water Pik), measured the withdrawal, and analyzed in detail the chemical and electrical changes in the neurons during this behavior—daily, weekly, for years. Though habituation may not be a sophisticated form of memory, Kandel did eventually discover the minute changes within the synapses, and the chemistry of cells that went along with the learning; it was the first time such a record of formation of memory taking place inside cells was made. Alkon took an additional step by working with a more full and robust form of memory than the withdrawal reflex: he explored *Hermissenda's* ability to learn a new style of life. Normally, the animals move toward light because, as they shinny along strands of seaweed, it is in the brighter, shallower water that they most often find their food. At the same time, the animals dislike agitation, and cling tightly when shaken, as when the seaweed they ride is buffeted too strongly by waves. Alkon determined to change these natural associations by shaking the animals while flashing lights, thus teaching the animals that light is a source of trouble as well as food.

Kandel and Alkon eventually revealed the chemical details of learning, which now plainly consists of complex little events that partake of both chemistry and electricity; learning occurs when neurons fire together in assemblies. In the simple animals, they saw for the first time the inner workings of the cells that constitute memory in all animals. After these visits, I began to imagine memory extended through history, from the early innovations of jellyfish to the complex creatures of the land which followed.

Memory is a sepulchre furnished
with a load of broken and
discarnate bones.
       —Joseph Glanvill, "The Vanity of Dogmatizing"

4. ⤳

The interval between our time in the sea and ourselves today seems unbridgeable. But I once did feel, in Africa, an immersion of time, three million years in a moment.

I had worked in Nairobi during the day. It was hot and the mission of a single interview had become a frustrating series of walks in the dust, taxi rides, and interminable waits among the warm gray metal of a slow and silent government. As evening came on, I gave up and drove out of the city, through the warm air, like a swimmer suddenly feeling the flow of the stream. I drove out into the grassland that lies very near the town, and after a time came to a great tree on a hill overlooking a small pool. In every direction there was nothing but tall, blond grass. I soon saw the animals moving, or rather the grass moving contrary to the wind. I watched as the animals warily assembled for water. I somehow felt completely comfortable here, suited to the elements in a way and with surroundings I had never before experienced, a memory coming up from the bones.

It was here that human memory took its long step beyond the purely animal. Our forebears, forest-dwelling apes, were forced from the trees by many years of dry and even drier weather. Twice in the past 5 million years it has happened as the earth has gotten chillier, and as a result more ice has accumulated in the glaciers. The trees gave way to grasslands, the great tall-grass savannahs of Africa. Our forebears had to foray more and longer into the grass. Once out of the trees, the posture of the pre-human apes of 4 million years ago became more vulnerable to attack and more dependent on signaling and social cooperation, including gathering and carrying. The gathering of food both necessitated and took advantage of upright walking. Up to that time, standing erect was simply too risky to do often—animals lost their cover and moved slowly in this posture, as two-legged running is much slower than four-legged. The pre-humans over time became upright scavengers, grouping into large families for defense. As the current belief goes, this transformation led to continuous upright walking.

The change of habitual posture required changes in body shape over time. The foot became streamlined and lost its hand-like grasping ability, the pelvis and thigh needed to shift shape and bulk to handle the load once before them, now on top of them. The changes radiated outward from the center of gravity and included ribs, respiratory organs, and the hands and arms that now became available and useful to manipulate and throw objects, among other things.

As these apes rocked back onto their pelvises and their backbones supported them, they did not immediately become tool-makers. Understanding of the step from forest-dwelling vegetarian to plains scavenger to nomadic hunter with a large brain and many tools is difficult; theories abound. The books speak of evolutionary change in which many markers seem to march in the same direction: brain size increases, tools appear, posture becomes upright, social groups become more elaborate, and so on. It appears now that the increase in brain size and perhaps tool-using as well was secondary to the shift to an upright posture.

One current line of research suggests that the expansion of

the brain occurred because of the need of the cranial tissues to spread themselves out for cooling. Once the tissues expanded for this reason, they were put to other uses. Other organs have similar histories, in which secondary purposes grew to dominate. The delicate, vibrating bones of the ear were first parts of the jaw and were displaced for other reasons to a position next to the ear canal. But these bones transferred sound well, and their natural reverberations were put to use to amplify sound. Similarly, light-sensitive cells eventually coagulated into retinae. Vocal areas of the base of the skull and top of the spine dropped over time to a lower position in the throat, allowing animals to breathe through not only noses but mouths as well, increasing the capacity to take in oxygen for extended exertion. This shift appears also to be the first requirement of uttering recognizable speech. It is the same tale as the Panda's thumb, which was a wrist bone recruited over time into the stripping of bamboo branches. Secondary use becomes primary, purpose is overturned. Perhaps something similar occurred in the growth of the mind.

In one line of research first reported in December 1994, it was found that the forebrain, where planning is carried out, was connected to and commandeered some of the work of the cerebellum, where body motion is controlled. Thus our movement in the world was echoed simultaneously in our mental images and our planning. The "mental space" that we now inhabit was created.

The drive of life is to survive and procreate, and for this the mind is the medium through which we take in the world, turn it over, and act upon it. We create within our biological selves, within our skins, a small, suited-to-ourselves model of the world and what to expect from it. Other creatures do the same, but in more limited ways. The dominating sense among amphibians is vision. Among reptiles and small mammals such as rats, smell is central to their being. In humans, none of the senses are extraordinarily developed, but several senses together provide a multichannel system of gaining information—vision, smell, hearing, and sensitive hands give touch a special significance as well.

Harry Jerison, a neurologist at U.C.L.A., says this marks the beginning of the evolution of mind. Combining the senses gives us the broad band of information needed to make a specially detailed mental map with which we can track and respond to the world. His idea is that the map creates a rich texture of thought and imagery that underlines the beginning of language.

I suspect mimicry was the first explicit language, and that imagery first made elaborate mimicry possible, in which we can picture a sequence of events and act it out for one another. Teaching by mimicry and entertainment by mimic-theater may have been the rudiments. This is how it is with children, and it is what makes theater especially resonant—that it is at the root of our evolution and our consciousness, before speech, before society, elemental. The mimicry was undoubtedly acccompanied by sounds that helped define and emphasize gestures.

In this way, we first began to share one another's consciousness. As Nicholas Humphrey says, social life among the primates is extraordinarily complicated. Forming and breaking alliances, manipulating others' behavior, keeping track of the status and relations of each member of the group and what that may portend in every encounter—these required the development of images and understandings into elaborate memories. "Like chess," Humphrey writes, "a social interaction is typically a transaction between social partners. One animal, may, for instance, wish by his own behavior to change the behavior of another, but since the second animal is himself reactive and intelligent the interaction soon becomes a two-way argument where each 'player' must be ready to change his tactics—and maybe his goals—as the game proceeds. Thus, over and above cognitive skills which are required merely to perceive the current state of play, the social gamesman, like the chess player, must be capable of a special sort of forward planning."

Thus, on the inner stage of memory, the actors of memory are moved about by the planners of the frontal lobes, creating a new dynamic set of memories, whole episodes from the world.

Roger Schank, a theorist of artificial intelligence, describes the way humans use these tools as they move about in the world using similar theatrical metaphors.

We operate day to day, in familiar situations, working from

what Dr. Schank calls "scripts," sequences of expectations built of previous instances. In a restaurant, we expect to be given menus, from which we choose food. We expect to pay at the end of the meal. We expect to find utensils, knife and fork in America and Europe, different utensils in other cultures. What is that bowl with water and lemon on the table? We must have a script for Asian restaurants or Middle Eastern restaurants to know, as finger bowls are no longer common in America.

"When we try to get a computer to understand English sentences . . . when we try to get a machine to reason from experience about medicine or law (as many expert programs have done), what we do is attempt to give that machine knowledge. We can't understand anything unless we know something else first. Then, understanding becomes a process of trying to relate what we are trying to understand to what we have already understood," Schank writes in *The Connoisseur's Guide to the Mind.*

"Scripts provide guides, but also create trouble," Schank notes. "Their power is that they tell us what is likely to come next. They help us make inferences without doing too much work. They package our expectations about restaurants and airplanes and department stores and allow us to operate effectively in new situations by drawing upon generalizations made from similar situations. Much of our ability to understand depends on scripts."

They can be elaborate. For example, the situations that develop in chess games are called "chess positions," and they provide great detail to the person who has spent hours building a sense of what is similar in the positions. One recalls the opening moves of a game which have gotten the label "Ruy Lopez opening," and the script provides likely moves to come next.

But the scripts also limit our thinking. "When we come to rely upon knowing how things are supposed to be, which event is supposed to follow which event, how people are supposed to act in certain situations, we have little ability to cope when things don't turn out as expected . . . People come to rely on situations where they know the scripts. Often they are afraid to venture into new arenas because they do not know the script and are afraid of looking foolish. The ability to learn depends on the ability to abandon scripts that are failing and to acquire new scripts," Schank writes.

For example, autistics have great trouble with novel situations, and find comfort in the repetition, often to an absurd degree, of familiar formulas of thought and action.

These fundamental developments were interior at first, and only gradually could be seen in the physical record of our tools, our campsites, our fire, our graves, and eventually our art.

The next turn beyond the scavenging tribes was, after the settling of weather after the ice ages, an era in which people began to cluster in larger and larger groups. What emerged, psychologist Julian Jaynes says, was the greatest of all cultural steps, the transformation of wandering tribes of twenty people to the settling of several hundred in the first town.

His example of an early town is Eynan, just north of the Sea of Galilee. It was created about eleven thousand years ago. "A town! Of course it is not impossible that one clan chief could dominate a few hundred people. But it would be a consuming task if such domination had to be through face-to-face encounters repeated every so often with each individual, as occurs in those primate groups that maintain strict hierarchies. I beg you to recall, as we try to picture the social life of Eynan, that these people were not conscious." Here he uses the term conscious to mean something more restricted than will; by conscious he means reflexively conscious, or self-conscious. The people of Eynan were conscious as other animals are, but not *self*-conscious as later humans are. "They could not narratize," as he puts it, meaning they could not form their mental images into cohesive stories. They had no "analog selves to 'see' themselves in relation to others."

The people of the city were "signal-bound," that is, they reacted each minute to cues, and were controlled by those cues. In that society, Jaynes says, the "cues" included oral language and imitation.

This was a time of transition for humans, when the rule of the chiefs began to be internalized. If a chief was not to meet and direct each individual, frequently and in person, some

mechanism must have developed for his rule to be felt and followed, without continuous instructions and decisions from the
head man. That is, people needed to internalize the chief's
words. In Jaynes's description, the people of the time literally
heard the king's voice as hallucinations intruding on their own
thoughts, guiding and aiding them. The creation of hallucinations, or elaborate memories, were animated mental models of
the chief and his sayings.

All this may not have occurred precisely as Jaynes proposes it,
but in some form it must have taken place. The passage into a
highly structured society must have made great demands upon
imagination and memory. Internalizing hierarchies and rules,
perhaps through the use of images of specific kings, who after
death must have spoken as gods, did occur in some way. Whatever the date of these events, the emergence of language, the
settling into fixed societies, and the creation of analog persons
and worlds within our minds must have occurred.

It was a new role for memory; it was more demanding and required the creation of internal "others." It was the invention of
story and a social passage of great import. Jaynes believes we can
see the origin of self-consciousness in ancient literature. The
earliest documents are instructions, as if dictated: orders, para-
bles, laws, recipes, lists—they are the substance of the most ancient writing. The people who appear in these documents are
impersonal, third parties. There is no "I" to be found, no description of reasoning or wading through feeling, no interior
monologues. From the earliest writings to the time of the Greek
philosophers, an important change has taken place in human
mentality. From the first versions of *Gilgamesh*, the ancient
Middle Eastern tale of material success and mortal woe, Jaynes
believes that its language changes from the stiff third-person de-
scriptive to the first-person interior narrative, from a document
with no "I" to one in which the great female goddess "speaks to
her own heart" of her feelings for the king Gilgamesh.

At some time, humans began to think of time as a "space" in
which people and things could exist as separate entities from
those we see passing before us in the moment. These people
and things can be brought to mind, re-experienced, regardless of

collective memory. [handwritten marginal note]

their presence or absence, closeness or distance. Only by translating the stream of time into a frozen field of space where things and events can be located can we create narratives, see ourselves as separate from our current thoughts, and manipulate that which is not present. Names for everything in this false-spatial field became both useful and eventually necessary. From existing animal-like as a momentary bundle of reactions in the present to placing ourselves on a mentally-created stage of all time and all action, finding our place upon it and seeing ourselves acting there—all this has come with the development of modern memory. The creation of humanlike memory raises us up from other animals, for better or worse.

5. ⤳

The earliest record of humanity is still written only in the land and its artifacts. But from the archaeologists' pits rise traces of human concern about memory: there are the cuts marked along the edge of a bone to keep track of the undulating phases of the moon; there are the great red-orange rubbings of elk and buffalo secreted inside deep caves and seen only by firelight; there are the flowers laid atop the bones of Neanderthals in shallow graves. These were the haunts of early memory.

But then, about ten thousand years ago, humans developed a peculiar habit of making marks upon stones and skins, clay and wax, and later birch bark in Russia, palm leaves in India, tortoise shells in China, and finally papyrus, parchment, and sheets of dried pulp paper. The distance between this crabbed bit of behavior and more sensible activities such as fighting and playing, eating and making love, is extraordinary. Watching humans emerge from evolution into writing is like watching the onset of insanity, in which our original simple purposes become detached from their usual sensible effects, and break free to drive new and seemingly strange behavior. For example, whacking one stone upon another, not once, but 50, 200, and 300 times to make something, and the odd behavior in which dark marks are laid out on stone or paper in sequences of thousands at a time. This became the greatest of the practical arts of memory, the creation of records themselves.

The invention of writing made memory manifest. Writing was created, not as a mystical or intellectual art, but rather as a clever marking system for commerce before its power as a medium of mass communication became clear. Denise Schmandt-Bessarat, scholar of ancient writing, has tracked the use of letter and picture symbols back to what appears to be its earliest incarnation: as markers to keep track of trade. The quantity and kind of products in a shipment—20 jars of wine, 20 jars of barley, 40 sheep—were impressed on the outside of clay purses that went with the deliverer of the goods. Inside the purse were small clay tokens representing the same count. When the goods and the purse were received, it was immediately possible to check the purse against the number of urns actually arriving. Soon symbols created for use in this way became independent of the delivery number, and were impressed on tablets as accounting records. The list of symbols grew until there were enough to use as a system of general representation, an alphabet. In this way we conquered the world, by naming and counting its parts, and assigning them places within our minds and in our archives. Flights of fancy and philosophy would not be far behind, but literature began with lists.

Before writing, ancient societies had designated certain officials to be the memory of society. In Greece, they were called mnemons; in Ireland they were the shanachie; in Inca culture they were the makers of the elaborate string records, the quipus, where everything was recorded from the census to the calendar. In each society they became the memory of the culture, they taught apprentices the arts of memory, and were themselves privileged citizens.

Great lists of persons and events had been routinely recited, and when writing began, these lists were set down once and for all: in the second section of the *Iliad*, French historian Jacques Le Goff writes, "we find in succession the catalogue of ships, then the catalogue of the greatest warriors and the best Achaian horses, and immediately afterward the catalogue of the Trojan army," with the references totaling about half of the entire second section. Still, writing seemed suspect at first; various officials in Egypt, Greece, and Rome said the employment of writing would result in flaccid memories. Said Plato, the Egyptian god Toth created writing, and Toth promised that it would make the Egyptians wiser and give them better memories; writing could be used for both memory and wit. But Plato, in the voice of Thamus, tells Toth, "The inventor of an art is not always the best judge of the utility or inutility of his own inventions . . . for this discovery of yours will create forgetfulness in the learners' souls, because they will not use their memories; they will trust to the external written characters and not remember of themselves . . . you give your disciples not truth, but only the semblance of truth; they will be hearers of many things and will have learned nothing; they will appear to be omniscient and will generally know nothing; they will be tiresome company, having the show of wisdom without the reality."

He said that it would be a fool who would leave behind only writings to convey his ideas, though that is all we now have of him. He preached, "I cannot help feeling that writing is unfortunately like painting; for the creations of the painter have the attitude of life, and yet if you ask them a question they preserve a solemn silence . . . and when [speeches] have been written down, they are tumbled about anywhere among those who may or may not understand them . . . and if they are maltreated or abused, they have no parent to protect them; and they cannot protect or defend themselves." In this statement is the kernel of the argument against free speech, overcome only two millennia later and only in some few places on the planet, and perhaps temporarily even there. The fear afflicting Plato also worried Julius Caesar, who refers to the young Gauls who would flock to the Druids for teaching: "There they learn by heart, people say, a large number of verses, and some of them spend twenty years studying with the Druids." The Druids, however, refused to write down what they taught: "They seem to have established this custom for two reasons: because they do not wish to divulge their doctrine, or to see their pupils neglect their memory by relying on writing, for it almost always happens that making use of texts has as its result decreased zeal for learning by heart and a diminution of memory."

Dismally, we find that it took no time from the creation of writing to arrive at a modern politics of memory, and methods of mass manipulation. Even in earliest days, the genealogists and memory-men were under the control of the chiefs, and in fact worked in the palaces for the most part. When writing began in earnest, it is no accident that most of what we know of the ancient world is in monuments to leaders—from the code of Hammurabi to the stelae of Egypt. It became, in fact, almost a comical exercise in Greece and Rome. Dr. Le Goff, director of studies at École des Haute Études en Sciences Sociales, recalled Louis Robert's description of Greece and Rome as "the civilizations of epigraphy" in which the inscriptions accumulated in temples, cemeteries, public squares and avenues, along roads, and "even deep in the mountains, in the greatest solitude." They encumbered the Greco-Roman world with an extraordinary effort of commemoration and perpetuation of memory. Le Goff says with a bit of vinegar on his tongue, "Stone, usually marble, served as a support for an overload of memory. These 'stone archives' added to the function of the archives proper the character of an insistent publicity, wagering on the ostentation and durability of this lapidary and marmoreal memory." The capital city thus became the center of the social mind, "the center of a politics of memory—but the king himself deploys, on the whole terrain over which he holds sway, a program of remembering of which he is the center."

There was a struggle between the personal and the collective memory; after all, a substantial portion of what memories we have of ourselves, our group, our nation, are held as shared ideas. They are shaped and referred to regularly by public figures. The notion of a Frenchman about an American is part of a pooled memory which may or may not conflict with the comparable notion we each hold personally. It is a common finding among poll takers that people simultaneously hold views shared with society and views contradicting the shared views.

The answer to the question of whether a citizen believes in

God or whether Roosevelt was a great leader will depend entirely on the frame in which the question is hung. Le Goff says that at an important metaphorical level, in the same way that amnesia is not merely a local disturbance of the individual's memory but causes more or less serious perturbations in his personality, the absence or loss of collective memory among peoples can cause serious problems of collective identity. One desperate example to which he refers is the Etruscans, who occupied Italy before the Roman Empire, and had there a flourishing and culturally sophisticated civilization. But alas, they kept all their history tightly bound—in the memories of the dominant social class alone. As the historian Mansuelli laments, we know the Etruscans only through the intermediary of the Greeks and Romans; no historical account, even if we admit that such an account existed, has come down to us. Perhaps their national historical traditions disappeared along with the aristocracy that seems to have been the repository of the moral, juridical, and religious patrimony of the nation. When the nation ceased to exist as an autonomous nation, Le Goff says, the Etruscans who still lived, carrying on with their shrinking culture and memories for some time, seem to have lost consciousness of their past, that is to say, of themselves.

Who controls the society's memory controls its will. The control of collective memory has been an important issue in the struggle for power among social forces. "To make themselves the master of memory and forgetfulness is one of the great preoccupations of the classes, groups, and individuals who have dominated and continue to dominate historical societies. The things forgotten or not mentioned by history reveal these mechanisms for the manipulation of collective memory. Accordingly, the first spate of human writing takes place under the control of kings and is heavily weighted toward praise of the regime, to the exclusion, often, of almost everything else," Le Goff writes. Kings constitute memory-institutions: libraries, archives, museums, as do presidents now.

Memory is the raw material of history, whether in the mind or scratched on papyrus. It is the living source from which historians draw. But memory and history are not the same. Memory itself is dangerously inaccurate, and this is compounded by the selectivity of historians among the ruins of memory. The race remembers first through individuals, then through records. Much is lost, and sometimes what is saved is nevertheless neglected by society: each generation learns what to remember anew. In ancient times, it was not so important that memory be faithful to an event or source, but that it be alive and made pertinent. At the end, the discipline of history not only draws from memory, but returns material to it, and thus moves the mental life of society, "the great dialectical process of memory and forgetting experienced by individuals and societies."

6. ⤤

In the earliest documents, we find that almost all references to memory are exhortations to remember wars, heroes, or the commandments of gods. In Homer, which may be placed in about 800 B.C., memory is the great prize for mortals, as they cannot attain immortal life after death.

But in early writing we also find a curious chasm developing, three thousand years ago. Here, the great split between East and West developed as it now is: the Hebrew and Christian works turn on commands, while the Chinese and Indian works on a subjunctive mood.

The Bible and its successors, the Koran and the book of Mormon, are dominated by commands to remember, to keep constantly in mind the divine rules. In the beginning, God offered a mnemonic to help in the effort. "And it shall come to pass, when I bring a cloud over the earth, that the rainbow shall be seen in the cloud: And I will remember my covenant, which [is] between me and you and every living creature of all flesh; and the waters shall no more become a flood to destroy all flesh. And the bow shall be in the cloud; and I will look upon it, that I may remember the everlasting covenant between God and every living creature of all flesh that [is] upon the earth. And God said unto Noah, This [is] the token of the covenant, which I have established between me and all flesh that [is] upon the earth."

And lest they forget: "If thou do at all forget the Lord thy God, and walk after other gods, and serve them, and worship them, I testify against you this day that ye shall surely perish."

By contrast, the holy books of Confucius, Buddha, and the Hellenistic world, later by a few centuries, were more philosophical, less hortatory and more advisory. We see Buddha instructing a follower on how to achieve a perfect holiness: "The disciple said: 'Can a humble monk, by sanctifying himself, acquire the talents of supernatural wisdom . . .' and the Blessed One replied: 'These are wondrous things; but verily, every man can attain them. Consider the abilities of thine own mind; thou wert born about two hundred leagues from here and canst thou not, in thy thought, in an instant travel to thy native place and remember the details of thy father's home? Seest thou not with thy mind's eye the roots of the tree which is shaken by the wind without being overthrown? Does not the collector of herbs see in his mental vision, whenever he pleases, any plant with its roots, its stem, its fruits, leaves, and even the uses to which it can be applied?'"

It is in the early philosophical-religious writings of the East, of China, and of Greece that memory first appears as a noun, a subject in itself. Not coincidentally, at the same time, the regular accumulation of knowledge began: in each subject of interest to the flourishing race, the simplest and most profound questions began to be asked and answers given. In physics, the questions were: Of what material is the world made? What is the smallest, most elemental unit, the *atomos*, of that worldly material? And how is it combined to give us the variegated face of the world which we see? By 50 B.C. Lucretius, in the lengthy poem *De Rerum Natura* (*On the Nature of Things*), laid out the idea of a scientific physics. He offered no method of study, and thus no hope of checking his assertions, but the work was begun.

When the powers of thought were turned inward and themselves became subjects for analysis, progress was more elusive. There was a curious blind spot. Physical features spring from one another, but where do thoughts spring from? A separation of the physical and mental began.

Plato was adrift in the romance of the mind, and believed that the nonmaterial regions of thought were the realm of the real. The perceived world was merely the medium through which thought moved, the dirty and uncomfortable sea in which pure bodies of thought floated. Physical and spiritual must be separate stuff, and this formed his view of memory.

On memory, Plato first tells us, "I would have you imagine, then, there exists in the mind of man a block of wax, which is of different sizes in different men; harder, moister, and having more or less purity in one than another, and in some of an intermediate quality . . . Let us say that this tablet is a gift of Memory (Mnemosyne), the mother of the Muses [the Muses are the nine properties of the imagination, and Memory begets them all]; and that when we wish to remember anything which we have seen, or heard, or thought in our own minds, we hold the wax to the perceptions and thoughts, and in that material receive the impression of them as real, the seal of a ring; and that we remember and know what is imprinted as long as the image lasts; but when the image is effaced, or cannot be taken, then we forget and do not know."

But later Plato drops this analogy for an ornithological one, in which he imagines the mind as an aviary, and the memories birds that we must chase down. The soul, caged inside the body, animates the brute body with life and thought, but is free to fly to metaphysical skies at death. We have understanding, he said, only when, by logical argument and experience, we shed our forgetfulness and listen to the singing of the soul. Further, he said, all knowledge must be simply recollection, because the soul is undying. Nothing could be learned, but only remembered from past lives and purer states.

Aristotle, his student, disagreed. He argued that nothing in the mind could come there but by the senses, or by mixing, after the senses had deposited them in the mind:

"All men by nature desire to know. An indication of this is the delight we take in our senses; for even apart from their usefulness they are loved for themselves; and above all others, the sense of sight." He said, "from memory, experience is produced in men; for the several memories of the same thing produce finally the capacity for a single experience.... science and art come to men through experience ... art arises when from many notions gained by experience one universal judgment about a class of objects is produced."

Curiously, at the same time Plato's and Aristotle's theories of the mind remained lively if vague, spiritually bound, and difficult to test, the practical arts of the mind were well advanced. Rhetoric was taught rigorously, and under this heading came many subjects, from politics and persuasion to methods of remembering large amounts of material for speeches and discussions.

The methods of enhancing memory were codified by 500 B.C., according to tradition, by Simonides of Ceos. The methods described, in one or another form, were used steadily in schools and by people of accomplishment for the next two thousand years, until the decline of memory in culture in the seventeenth century.

The classes on rhetoric might be thought of as combined courses of literature, law, political science, history, and government, subsumed under a single title. A practical problem was at the center of this concentration: very little of knowledge was written down. Most was passed orally, and day-to-day government had to be conducted from the front of the mind, without aids. Thus knowledge and memory were closely allied.

"Memory ... serves as a universal treasure house, and it has to be given safekeeping of every single aspect of the speech one is going to make," wrote Cicero, "all its substance and all its words." Without memorization, "however remarkable all these items were when the orator originally conceived them, they will, one and all, be totally wasted."

The method used and taught by Cicero, and which had been current for five hundred years already, was still essentially that of Simonides. The tale of this man Simonides, a poet who was the first to charge for his work and a speaker of great repute, has been told by historian Frances A. Yates, who herself is perhaps the only historian of memory. She writes that "At a banquet given by a nobleman of Thessaly named Scopas, the poet Simonides of Ceos chanted a lyric poem in honor of his host but including a passage in praise of Castor and Pollux. Scopas meanly told the poet that he would pay him half the sum agreed upon for the panegyric and that he must obtain the balance from the twin gods to whom he had devoted half the poem. A little later, a message was brought in to Simonides that two young men were waiting outside who wished to see him. He rose from the banquet and went out but could find no one. During his absence, the roof of the banqueting hall fell in, crushing Scopas and all the guests to death beneath the ruins; the corpses were so mangled that the relatives who came to take them away for burial were unable to identify them. But Simonides remembered the places at which they had been sitting at the table and was therefore able to indicate to the relatives which were their dead. The invisible callers, Castor and Pollux, had handsomely paid for their share in the panegyric by drawing Simonides away from the banquet just before the crash. And this experience suggested to the poet the principles of the art of memory . . . Noting that it was through his memory of the places at which the guests had been sitting that he had been able to identify the bodies, he realized that orderly arrangement is essential for good memory."

The memory operates by association, and so it is easier to recall a whole sequence of information if it pairs with a sequence of things we know for certain: for example, the layout of our own homes. This became the method of the "places," or in the Latin, the method of the "*loci.*" The rememberer was to memorize material by associating each part to be remembered with a place. Most commonly, it was suggested he walk through his own house, or a familiar square, and assign to each room or each outstanding feature one of the items to be remembered. In the vestibule, one may leave an image of a dead, bloody soldier to recall the introductory thought about the war dead, and on the wall are two pictures, each of which may have another striking or extraordinary image that would call to mind another point to be remembered.

Speeches from that day have thus used the phrases "in the first place" and "in the second place," as the speaker walked through his oration and his home at the same time.

The method plays powerfully on the idea of association,

which, if philosophers could have examined it closely at the time, would have yielded far more information about the nature of memory than was gained by the abstract reasoning popular at the time. Books on the improvement of memory are still published in a steady stream—for example, the books by Harry Lorayne, or a more recent one by Peter Russell—still using the method in one shape or another.

Until recently, however, psychologists, for the most part, had become uninterested in such mnemonic methods. Karl Pribram described the reintroduction of the method to an unwary psychologist:

"The antagonistic attitude of experimental psychologists toward mnemonic devices is even more violent than their attitude toward their subject's word associations. Mnemonic devices are immoral tricks suitable only for Gypsies and stage magicians. . . . One evening we were entertaining a visiting colleague, a social psychologist of broad interests, and our discussion turned to Plans [methods of association and remembering].

" 'But exactly what is a plan?' he asked. 'How can you say that memorizing depends on plans?'

" 'We'll show you,' we replied. 'Here is a plan that you can use for memorizing. Remember first that' ":

> one is a bun,
> two is a shoe,
> three is a tree,
> four is a door,
> five is a hive,
> six are sticks,
> seven is heaven,
> eight is a gate,
> nine is a line, and
> ten is a hen.

" 'You know, even though it is only ten-thirty here, my watch says one-thirty. I'm really tired and I'm sure I'll ruin your experiment.'

" 'Don't worry, we have no real stake in it.' We tightened our grip on his lapel. 'Just relax and remember the rhyme. Now you have part of the Plan. The second part works like this: when we tell you a word, you must form a ludicrous or bizarre association with the first word in your list [as in the method of the *loci*, using places], and so on with the ten words we recite to you.'

" 'Really, you know, it'll never work. I'm awfully tired,' he replied.

" 'Have no fear,' we answered, 'just remember the rhyme and then form the association. Here are the words' ":

1. ashtray
2. firewood
3. picture
4. cigarette
5. table
6. matchbook
7. glass
8. lamp
9. shoe
10. phonograph

"The words were read one at a time, and after reading the word, we waited until he announced that he had the association. It took about five seconds on the average to form the connection. After the seventh word, he said that he was sure the first six were already forgotten. But we persevered. After one trial through the list, we waited a minute or two so that he could collect himself, and ask any questions that came to mind. Then we said, 'What is number eight?'

"He stared blankly, and then a smile crossed his face, 'I'll be damned,' he said, 'it's lamp.'

" 'And what number is cigarette?' He laughed outright now and gave the correct answer. . . . We proceeded to demonstrate that he could in fact name every word correctly."

There are limits to the technique suggest by Pribram, as one tale I found in Edmund Fuller's dictionary of anecdotes conveys:

"Memory training by association became a fad in a certain school. 'For instance,' the English teacher was explaining, 'if you want to remember the name of a poet, Bobbie Burns, you might conjure up in your eye a picture of a London policeman in flames. You see, 'Bobbie Burns.'

" 'I see,' said one of his pupils, 'but how is one to be sure it doesn't represent Robert Browning?' "

The arts of memory were revived and rewritten periodically by one scholar after another through the centuries. Cicero wrote out a version of the method for orators, but he confessed:

"For my part I wonder at memory. For what is it that enables us to remember, or what character has it, or what is its origin? I am not enquiring into the powers of memory which, it is said, Simonides possessed . . . I am speaking of the average memory of man."

Cicero said average memory is responsible for all the great inventions: first, assigning of names to everything; next, gathering people together in a social unit; then, the first writing; and later the discovery of the revolution of the heavens; and so on. Cicero continues: "A power able to bring about such a number of important results is to my mind wholly divine. For what is the memory of things and words? What further is invention? Assuredly nothing can be found even in God of greater value than memory."

Part II

How great is the world in the light of lamps!
How small is the world in the eyes of Memory!
    —Charles Baudelaire, *Les Fleurs du Mal*, 1857

7.

I know of a fervent young man who, intelligent but uncertain where to lay his belief, went to the Franciscan Friars with the thought of being a monk and teacher in the humble tradition which runs back to pre-Christian times. He went to the monastery, awoke just after 4 A.M., knelt on stone in the chapel to serve at Mass, took his breakfast cold, sang the mesmerizing Gregorian chants of the morning, and went to the field for a day's labor with the brown-robed and hooded figures.

He was thrilled with the simplicity and beauty of it, but expected that in the evening there would be reading and scholarly discussion, debate about mysteries. He was disappointed; the amiable friars could not talk freely about belief and God and doctrine; they seemed puzzled when the questions became difficult, and sought comfort in the repetition of doctrinal sayings. The young man left eventually, disturbed by the peace, peeved with the men who had no ambition but to succor in the bosom of the Church. In the great institutions of history, individuals have often found comfort, but not expression.

This young man's experience put me in mind of Aurelius Augustine, who was one of those whose personal tale was vivid and extraordinary, but who was split between personal feeling and institutional belief. He was a North African, a Berber who was raised under Roman tutelage, in a town near Carthage. His father, though not Roman, rose in the local government to a substantial position and married a woman who had become an avid Christian.

Augustine was developing an interesting scholarly career in North Africa when the first of two great blows in his life came. He was a passionate man; a friend he met at school became very close to him. He was an intellectual companion and fast friend of the heart, who died suddenly when both were young men. He was "sweet to me above every sweetness of my life," Augustine wrote in his *Confessions.* "My heart was made dark by sorrow and whatever I looked upon was death. My native place was a torment to me, and my father's house was a strange unhappiness. Whatever I had done together with my friend, was, apart from him, turned into a cruel torture. My eyes sought for him on every side . . . Therefore I raged and sighed and wept and became distraught, and there was for me neither rest nor reason . . . Where could my heart fly to, away from my heart? Where could I fly to, apart from my own self? Where would I not pursue myself? But still I fled from my native town."

He went to Rome, where he fell in with devout Christians whose solace he was drawn to. But they demanded something of him before he could join their strict, celibate ranks. In Africa he had fallen in love with a woman who, he said, was too far below his station for the two to be married. But he stayed with her for thirteen years, she bore him a son, and she had joined him when he fled to Italy. Augustine said there was great happiness in his domestic life, but his devout mother and his new friends pressed him toward purity rather than happiness—toward marriage, in fact, but not marriage to his "unworthy" partner. His mother and his friends selected for him a young Christian woman to set him aright. Then, to complete the act, his friends kidnapped his lover and sent her back to Africa.

Augustine wrote: "The woman with whom I shared my bed was torn from my side as a hindrance to my marriage. My heart which clung unto her was torn, wounded and bleeding. . . . It would be two years before I could marry. Not yet healed was the wound which had been made by the cutting away of my companion, but after inflammation, and most acute pain, it became a scar, and my pains became less acute, though more desperate."

From his pain he fled deeper into religion, and still in hurt, fled to memory. Hounded by grief, he fell into a severe depression, as we now can see with modern eyes. In this state, Augustine found that even the loveliest of memories could worsen his mood. He became the first voice of memory and reflection, the first appearance of a fully fleshed person in the literature of memory, who carried both delights and horrors for memories.

He felt any pleasure must be sin, and became determined to avoid all pleasures: the taste and feel of food when he was hungry, the smell of flowers, the sound of music. "When it happens that I am moved more by the singing than what is sung, I confess that I have sinned, in a way that deserves punishment," he wrote. And the pleasures of the eyes! "The eyes love fair and varied forms and bright and beauteous colors. Let not such things possess my soul."

Light made him feel glad, and he hid from it. In this deprivation, he had left only the furniture of his own mind. He devoted hours to it, thinking about memory, puzzling out how it is formed, and what it contains. In Shakespeare's phrase, the graves of the memory rendered up their dead for him. He wrote, beginning in 401 A.D., in the most extended and perceptive passage on memory in human literature to that time. He went inward, as he thought, toward God. And there he came upon "the fields and spacious palaces of my memory, where are the treasures of innumerable images, brought into it from things of all sorts perceived by the senses. Hidden away in that place is whatever we think about . . . When I enter there, I ask whatever I want brought forth, and something instantly comes; other things may be longer in coming, which are fetched, as it were, out of some inner receptacle; still others rush out in troops, and while one thing is desired and required, they start forth, each asking me, 'Is it perchance I?' These I drive away with the hand of my heart, from the face of my remembrance; until what I wish for be unveiled, and appears in sight, out of its secret place . . ."

"There all things are preserved distinctly, each having entered by its own avenue: as light, and all colors and forms of bodies by the eyes; by the ears all sorts of sounds; all smells by the avenue of the nostrils; all tastes by the mouth; and by the sensation of the whole body, what is hard or soft; hot or cold; smooth or rugged; heavy or light; either outwardly or inwardly to the body. All these the great harbor of the memory receives in her numberless secret, and inexpressible windings. Each is brought out at need; each enters by his own gate, and is there laid up." He was, says modern scholarship, more correct in this than he could know: memories which come in by the eye are ultimately stored in the visual regions of the brain, while the auditory fragments of the memory are harbored in the aural areas.

Augustine gave himself over to memory, and believed he would find God there. His prayers were pleas: "What sayest Thou to me? See, I am mounting up through my mind toward Thee who abidest above me. Yea, I now will pass beyond this power of mine which is called memory, desirous to arrive at Thee." Exercise memory and will though he would, his anguish never completely left him.

I have visited Rome a number of times myself, and have found that the sharpest pleasure in the visits is a casual feature of the city, which must have been present already in the time of Augustine. It is a haphazard mix of time in the streets, parks, and buildings, objects discarded randomly from nearly three thousand years of human habitation. At a bus stop once I found myself sitting on what I thought was a bench; as I got up I realized it was a worn piece of Roman column, merely debris. Across the piazza from that was a fifteenth-century building, shoulder to shoulder with one of the eighteenth century and another of the twentieth century. The place of which I am fondest in the stew of Roman times is a short walk up from the Tiber River and just across the bridge from the palaces of the Vatican and St. Peter's, the Campo di Fiore, field of flowers, which has by an accumulation of accidents become a square that venerates freedom of speech. Here the earliest working printing press in Italy operated when not suppressed, and here is the sculpture of the great Giordano Bruno.

After Augustine had come a millennium of intellectual stagnation in Europe, but one of the figures who marks the return of light was Bruno in Italy. He was, apart from other things, a physicist, among the greatest of the Renaissance, and a philosopher who believed that there was not one but many ways to view the world. Absolute truth, he said, should not be the beginning of knowledge and may not be possible in any case. He defended, and elaborated on, Copernicus's world system, in which the sun holds the center.

He also wrote several books on memory, and was one of those who preached a new vision based on the ancient arts of memory. These arts were hidden until the revival of science and scholarship in the Renaissance, when the whole of ancient thought was raised like a sunken ship from the depths of time. The old rules of rhetoric from Greece and Rome returned, as did the exaltation of memory and an enthusiasm for the power of the human mind. New texts on memory were written, new theories of knowledge devised. The greatest literary work of the time, the *Divine Comedy* of Dante, was itself patterned on, and mentions within it, the arts of memory and their organization.

Bruno, Dante, and others raised a glorious possibility: that a man, if he were bright and brave enough, could use certain techniques to set his mind in synchrony with the way the universe is organized. Thereby, he could just consult his memory to achieve new knowledge. It was known, after all, that memory worked by association—recalling one or another detail might draw up a third item out of memory. It followed that if the human memory could be organized meticulously upon a model of the world itself, we might explore the memory and discover, by association from facts we know, facts we do not yet know.

Enthusiasm was stirred by this idea and by the plans for the device called the Memory Theater, in which the Italian Guilio Camillo put a complete description of knowledge to that time—organized, as the universe was, by planetary signs under which areas of knowledge were written as stories. Observers were to stand onstage, and look out into the theater, where, along the aisle of Mars, one would find descriptions of the tactics of war. Under Saturn would appear the elements of the Earth, and under Mercury the activity of those elements; hence geological and agricultural knowledge would be at the crossing of these two.

"This may be more clearly expressed," Camillo wrote, "from the following illustration. If we were to find ourselves in a vast forest and desired to see its whole extent, we should not be able to do this from our position within it, for the view would be limited to only a small part of it by the immediately surrounding trees, which would prevent us from seeing the distant view. But if, near to this forest, there were a slope leading up to a high hill, on coming out of the forest and ascending the slope we should begin to see a large part of the form of the forest and from the top of the hill we should see the whole of this. The wood is our inferior world; the slope is the heavens; this is the supercelestial world. And in order to understand things of the lower world it is necessary to ascend to superior things, from whence, looking down from on high, we may have a more certain knowledge of the inferior things." Historian Frances Yates comments that as "one thinks of all these drawers or coffers in the Theater, it begins to look like a highly ornamental filing cabinet. But this is to lose sight of the grandeur of the Idea—the Idea of a memory organically geared to the universe."

The noble families of Italy called Camillo to their *palazzi* to explain this marvelous system of remembering and knowing, and the king of France brought Camillo to his court for extensive talks. The king contributed funds for the building of a full-sized Theater of Memory, and it seemed all knowledge was within their reach. But alas, Camillo never got farther than drawings, of which we have a bare sketch, and small models, which have been lost. (In this, Camillo's experience was remarkably similar to that of Charles Babbage, the English creator of a great computing engine in the nineteenth century. It was to be a giant mechanical rather than electronic computing machine, made with the steam, polished brass, and wood of the last century. Babbage got a grant, but never finished the machine, though scholars now tell us the huge computer would have worked, and perhaps if Babbage were luckier, the computer could have entered society a century earlier than it did.) As for poor Bruno, memory was his death. Once, while he was out of Italy hiding from the pope, who wanted to prosecute him for his cosmological heresies, he was lured back by his vanity in these matters. An invitation from an Italian family to lecture about theories of memory brought him across the border. At once, he was arrested, and eventually burned in a pyre on the northeast corner of Campo di Fiore, a few yards from where his cloaked figure now stands.

I cannot paint what then I was.
The sounding cataract
Haunted me like a passion: the tall rock,
The mountain, and the deep and gloomy wood,
Their colors and their forms, were then to me
An appetite; a feeling and a love,
That had no need of remoter charm,
By thought supplied, nor any interest
Unborrowed from the eye.—That time is past,
And all its aching joys are now no more . . .
                    —Wordsworth, "Tintern Abbey"

8. ᪣

Giordano Bruno's liberal and expansive notion of the universe flourished after him, as did his hope for the methods of memory, and the integration within it of all knowledge. Among his champions a century later was Baron Gottfried Wilhelm von Leibniz, one of two creators of the calculus and modern mathematics. In his work, and that of his competitor, René Descartes, we read again of memory systems and the possibility of learning the great secrets of knowledge by seeing clearly and deeply into our own memories. Descartes, after perusing the notions, said finally that memory systems would not succeed as the Renaissance dreamers had hoped, but Leibniz still believed. Leibniz wrote that it would be the potent mix of memory arts and the calculus which could be the realization of Camillo's and Bruno's schemes to map all knowledge. Leibniz wrote of a great encyclopedia of all knowledge, in which "characters" could be assigned to each and every idea.

And so, says Frances Yates in her history *The Art of Memory*, the "universal calculus" would be established to solve all problems. Leibniz had visions of the application of the calculus to all departments of thought; even religious conflict would be dissolved by it. Dr. Yates writes "Those in disagreement, for example, about the Council of Trent, would no longer go to war, but would sit down together, saying, 'Let us calculate.'"

But just as these flights of intellectual fancy rose, the burdens of the time brought them down again. The era was, after all, the end of rational science and the beginning of empirical science; one must experiment, not only reason, and with the advent of experiment came its great river of data—facts, lists, observations, calculations, in greater and greater flood. Memory soon overflowed its bounds. More and more urgently, memory had to be deposited outside human minds, outside the reach of even the most prodigious human memory. It is not an accident that practical printing methods were created at this time, and manuscripts multiplied to contain the tide of knowledge.

Descartes saw this, and wrote of the old systems of memory that they were too much work to be sustained. Instead, he took up the method of science, experimentation, and turned it on the subject of memory itself, as well as other features of the mind. A devout Christian, he tells us that he abandoned all traditional ideas (well, almost) deliberately in order to put knowledge on a foundation of empirical proof.

He wrote of his doubt, "so serious are the doubts into which I have been thrown . . . that I can neither put them out of my mind nor see any way of resolving them. It feels as if I have fallen unexpectedly into a deep whirlpool which tumbles me around so that I can neither stand on the bottom nor swim up to the top." He said he wished to substitute proof and observation for scholastic authority, yet repudiate skepticism concerning the existence of God and the soul, writes Stephen Priest in his *Theories of the Mind*. Descartes's conviction was that, if some item of knowledge could not be doubted, then that item was absolutely certain, and the rest of knowledge could be reinstated using that certain knowledge as a first premise.

Notoriously, Descartes doubted in turn the evidence of the

five senses, the existence of physical objects, the truths of the various sciences, the existence of God, the claims of mathematics and geometry, and all the various kinds of truth he had previously taken for granted. This procedure culminated in his being unable to cast doubt on just one belief—the belief in his own existence. He concluded that, just so long as he doubted, he had to exist in order to do the doubting. But when it came to building the world back up from his first principle, his imagination failed him. He could not create an image of humans without an aperture through which the mind reached into the realm of pure reason, to God. He thus declared that while both body and mind were substances, the body was "extended" substance, *res extensa,* with size and heft, while the mind was substance without observable extension in the world, *res cogitans.* He suggested that the opening to God and the rational from the palpably physical machinery of the brain was the pineal gland, an organ about the size of a garden pea and located near the back center of the brain. This was a central difficulty in the study of mind and brain from ancient times to the present: the trouble of imagining how the apparently undifferentiated mass of the wet brain could give rise to the fine spirits of thought.

The instruments of science were not idle, however, as philosophy sputtered. By the middle of the eighteenth century, Luigi Galvani in Bologna had stirred the intellectual world when he found that the muscles of a frog's leg contract when touched with charged metal rods. Galvani felt that the body created this force, electricity, and the belief that electricity was the vital force, the animating energy of life, swept Europe (and of course later created the *Frankenstein* story). To this conjecture was added a striking idea by Paul Broca of the Bicetre Hospital in Paris, and by the work of Carl Wernicke in Germany. They began to work with patients who (like Mr. M. later) had odd and distinct disabilities. Broca had questioned a patient who had lost his power to speak but who, upon examination, had a normal voice box. His comprehension was normal. The man, Monsieur Leborgne, was known to the lab workers as Monsieur "Tan" because that was the only word he had been heard to pronounce over the previous twenty years. In the spring of 1861, the man died, and Broca showed on autopsy that Monsieur Tan's inability to talk was caused by an injury, a single hole really, at the front of the brain just above the temple. Here was the first proof that a higher mental ability could be traced to a specific piece of tissue, to the meat of the brain. Descartes's substance of the mind was no longer unextended.

Soon, vision, hearing, and smell in animals each were located experimentally by the arts of surgery, and human patients with peculiar mental disabilities began to be examined for anatomical damage, for indications of how they might compare to those with normal human perception and thought.

Still, peering down through the tube of a microscope at the material of the brain was disappointing. It did not yet seem possible that thought could be carried on by what the scientists saw there. At best, when the brain cells were immersed in dark stains to try to bring out their outlines, what emerged was an impenetrable forest of fine threads, vast in numbers and hopelessly tangled. The formula of René Descartes in these matters sufficed for two centuries. His belief that spirit still must be separate and thought must stand outside the physical remained a firm though shaken fashion of thought.

9. ↫

It was after science had taken root in the world of knowledge that a sort of second renaissance began. The start of it may be dated between 1880 and 1920, and it represented the realization, not of true appearances in nature, which those in the Renaissance sought, but of the true structure which lay beneath appearances. The revelation of atomic structure is perhaps the signature discovery of this second renaissance, as it speaks of wild motion and tiny beings active beyond vision, moving things which are nevertheless solid foundations for all we see.

Finding the reality beneath appearances became a theme of the age. Michaelson, Einstein, Bohr, and others laid out the frame in physics. In mathematics and philosophy, it could be said that Kurt Godel and Bertrand Russell sketched the path forward. In psychology both Freud and Watson, working at odds but both seeking deeper structure, altered psychology permanently. It may be seen mostly clearly now in art, beginning with the blistering of the canvas in impressionism and continuing to the total breakup of form and surface in the Fauvists and Cubists. Breakup of color can be seen in Monet's *Haystacks* in 1891 and Seurat's *Sunday Afternoon on the Island of La Grande Jatte*, 1884–86, and the breakup of form can be seen to start in

Cézanne's *Mont Saint-Victoire* of 1904, Braque's *Man with a Guitar* in 1911, and Duchamp's *Nude Descending a Staircase* in 1912.

The sudden advance of arts joined, or was the result of, the advances of science. Among the works of the same era were: *Das Kapital* and *Mein Kampf, Ulysses* and *Metamorphosis, Huckleberry Finn* and *Sherlock Holmes*. It was the time of *Dr. Jekyll and Mr. Hyde, The War of the Worlds,* as well as Eliot's "Prufrock," Akutagawa's *Rashomon,* and Synge's *Playboy of the Western World.* Modern and more had arrived with the discovery of atomic structure, the speed of light, and relativity. There was as well the foundation of genetics and the discovery of the substances of life, DNA and RNA.

The breadth of discovery was extraordinary, from the creation of the Kodak box camera and the motion picture film to the electrification of city and countryside. In philosophy, the *Zarathustra* of Nietzsche was written, and the works both mathematical and philosophical of Bertrand Russell; and the first book of Wittgenstein was published. In transport, there came autos and airplanes. There was also the creation of ragtime, jazz, and the codification of blues. We heard Schoenberg and Bartók, Elgar and Ives. There was Strauss and Sousa, the *Wizard of Oz* and Isadora Duncan, Gilbert and Sullivan, and the first roller coaster. Curiously, even religion was revived as well, though perhaps in reverse; this was the time when "the fundamentals" of Christian faith, an actual list of indisputable biblical principles, were laid down in horrified reaction to the ferment of the intellect.

In the study of memory, the list of investigators who sought deeper understandings included William James, Theodule Ribot, Hermann Ebbinghaus, and Sergei Korsakov, among others. A substantial work of philosophy, Henri Bergson's, was founded on an analysis of mental behavior and memory.

It is at this same moment that the first patients called amnesics appeared, described in the notes of Dr. Ribot in Paris.

With hopes of illuminating the usual processes of memory, Dr. Ribot studied the strange mishaps of memory which doctors had told one another as amusing stories for many years. In his book, *Les Maladies de la Memoire,* he said hopefully, "The disorders and maladies of this faculty, when classified and properly interpreted, are no longer to be regarded as a collection of amusing anecdotes of only passing interest. They will be found to be regulated by certain laws which constitute the very basis of memory, and from which its mechanism is easily laid bare."

Not so easily, perhaps. Ribot classed cases according to their depth and duration, according to similarities such as the inability to remember the names of objects. He compiled tales of amnesics, the first known kindred to Mr. M.

He told of the Englishman who descended into a mine where there were noxious fumes; he soon utterly lost the ability to speak German, in which he was previously fluent. Upon return to the surface, after some confusing moments of rest, he regained the language.

Ribot told of another, "An educated man, thirty-one years of age, who found himself at his desk feeling rather confused, but not otherwise ill. He remembered having ordered his dinner, but not eating or paying for it. He returned to the dining room, learned that he had both eaten and paid, showed no signs of being ill, and had set out for his office. Unconsciousness lasted about three-quarters of an hour."

And further: "A shoemaker, seized with epileptic mania on his wedding day, killed his father-in-law with a blow from his knife. Coming to himself at the end of several days, he had not the slightest consciousness of what had taken place."

Perhaps the most fascinating was "a young married woman, about twenty-four years of age, who, for about six weeks, remained in a continuous state of somnolence, the torpor increasing from day to day until finally (about the 10th of June) it became impossible to rouse her. She remained in this condition for nearly two months . . . Toward the end of August she returned, little by little, to a normal condition. On her recovery from the torpor, she appeared to have forgotten nearly all her previous knowledge; everything seemed new to her, and she did not recognize a single individual—not even her nearest relatives. In her behavior she was restless and inattentive, but . . . she was delighted with everything she saw or heard and altogether resembled a child more than a grown person."

Haunting moments, and haunted lives. Ribot described how

memory loss can pierce our sense of solidity, can invade our belief that we are who we are, that all goes on as before.

As I felt my way through this history, I spent the summer at the Marine Biological Laboratory. (It, too, was founded in this same period, 1888!) There in the musty stacks of the Lilly Library I read memory studies, intertwined with the growth of experiment in psychology. Physiologist Wilhelm Wundt in Leipzig established his laboratory of psychology founded on rationalism, and he studied the logic of thought in rigorous, timed experiments, designed to find the atoms of introspective thought. Memory was among William James's chief concerns, and was the first *exempla gratia* in the first chapter of his greatest work, *Principles of Psychology*. Sergei Korsakov in Russia at the same time found and delineated the most common of amnesic syndromes, the loss of memory experienced by alcoholics who have drunk so much for so long that great tracts of cells in their brains have been destroyed. The rush of discovery, all within the space of a few years, was referred to at the time as "the new psychology"; experimentation was the word of the day.

Among the most brilliant of those confronting the newly found issue of memory was Hermann Ebbinghaus. He took himself as a subject, and, unlike current psychologists, pulled himself through the tortures of his own experiment, studying daily for more than two years long lists of nonsense syllables. How rapidly could he acquire these pristine, seemingly meaning-free bits of language? And how soon would he forget them? Could he relearn them? He timed himself in learning and recall day upon day. It was the first scientific study of the psychology of memory, the first time a mental ability had been so carefully tracked and mapped.

What resulted was the beautiful if disturbing "curve of forgetting." It stands today, delineating the news that the greatest part of what we learn is lost in the first half-hour afterward, and the rest continues to decay with a slow certainty over weeks, months, and years, except for the smallest scraps. William James was impressed and called the first experiments of the new scientific psychology "a really heroic series of daily observations."

In Pavia, Italy, the anatomist Camillo Golgi found that the

cells of the nervous systems responded very oddly to silver—a few, but not all, absorb it, and once they do take it in, they distribute the substance through their full length and down all descending branches of the cell. The silver-filled cells stand out as thin black trees, sharp against the light background of unmarked cells. These Golgi studied and drew in careful ink portraits.

But the champion of these painstaking drawings was Santiago Ramón y Cajal, who adopted Golgi's method in Madrid, and began to understand much of what he was seeing. "Against a clear background stood black threadlets, some slender and smooth, some thick and thorny, in a pattern punctuated by small dense spots, stellate or fusiform. All was sharp as a sketch with Chinese ink on transparent Japan-paper. And to think that was the same tissue which when stained with carmine or logwood left the eye in a tangled thicket where sight may stare and grope forever fruitlessly, baffled in its effort to unravel confusion."

Ramón y Cajal found that the cells were each joined to others at tiny knobs, which came to be called synapses, after the Greek for "to clasp." The cells clasped not just one other cell, but many. Each was as a tree with a great number of branches at one end, a central body, and a fine array of roots at the other. At the branch end, the trees accepted information from many other cells; in the central trunk, they transformed the information into their own summary version of the signal, and then out through thousands of rootlets sent the new signal to the waiting branches of other cells.

The cells themselves, the basic units of mental life, were called neurons, from the Greek for string or sinew—the sinews of thought.

Pursuing an accurate image of these, Ramón y Cajal said, was like being the first person to enter a hidden garden:

"Like the entomologist in pursuit of brightly colored butterflies, my attention hunted, in the flower garden of the gray matter, cells with delicate and elegant forms, the mysterious butterflies of the soul, the beating of whose wings may some day—who knows?—clarify the secret of mental life."

"The nervous tissue contains the most charming attractions.

Is there in our parks any tree more elegant and luxuriant than the Purkinje cell of the cerebellum, or the psychic cell that is the famous cerebral pyramid?"

The brain had seemed to Descartes, and to many for years afterward, too simple and gelatinous an organ to be an interesting prospect for the seat of all thought. It had been rather like the moment in biology when the true nature of DNA was found. As late as 1953, DNA was thought to be a "stupid molecule," too simple and monotonously repetitious to be a prospect for the secret of life. It is only in the past few decades that the subtle power possessed by the brain's structure has been seen more clearly. It is complexity distilled. As far as we understand these matters, the brain is the densest, most complex object in the universe. The numbers alone speak of this: there are a hundred billion cells in the brain—twenty-five times the human population of the planet. Each cell seeks out and connects itself to others of its kind—a thousand to a hundred thousand junctions for each of the hundred billion cells. Further, at each of the thousands of junctions made by each of the hundred billion cells, there are numerous varieties of molecules to convey signals from one slender cell to another, and a variety of molecular hands to receive the chemical signals. Here is all the depth and diversity needed as the ground from which the mind may spring.

Ramón y Cajal once described listening to a Spanish leader speak of his deep religious feelings. The man told him that he did well to study the body's cells because they sometimes "influence" our actions.

" 'No,' I should have replied if respect and veneration had not sealed my lips. 'Those tiny cells, which keep hidden in their minuteness the mystery of life, are the whole man, in his two aspects, rational and physiological . . . they react to the stimuli of the environment, give us the illusion of free will, and, in fine, perform our actions in their completeness.' "

"All that we apprehend of the external world is brought to our consciousness," wrote another of the founders of neuroscience in the same era, Hermann von Helmholtz, "by means of certain changes which are produced in our organs of sense by external impressions, and transmitted to the brain by the nerves. It is in the brain that these impressions first become conscious sensations, and are combined so as to produce our conceptions of surrounding objects."

By the beginning of the twentieth century, in 1904, William James had restated the matter of mind for the current era. In his essay "Does Consciousness Exist?" he wrote that "There is . . . no aboriginal stuff or quality of being, contrasted with that of which material objects are made, out of which our thoughts of them are made." Mind, including thought and will, arises out of the physical order within the brain, and not from an infused soul. Mind stands not outside it, but within it. As flight is ascribed to the bird, thought is said to be a gesture of the brain.

James, the genial brilliance of the age, spent a life absorbing perspective in endeavors from the quivering tables of spiritists to the sudden descent into the unconscious of Sigmund Freud, whom he read and went to meet, to the most rigorous Germanic experiments on the timing of nerve cells.

James expected that the work of a science of psychology would seek to understand states of the brain and link them to the states of mind we experience. But psychology in 1900 was at the state similar to physics before Galileo had enunciated the laws of motion, to chemistry before Lavoisier had stated the laws of preservation of mass. James, characteristically, did not try to wrestle to the ground individual shards of knowledge about memory or mind at the time, but instead summarized grandly all that came to his attention, from Golgi to Ribot to Ebbinghaus. He became most absorbed in the flow of consciousness, and created the term "stream-of-consciousness" writing. "Consciousness, then, does not appear to itself chopped up in bits. Such words as 'chain' or 'train' do not describe it fitly as it presents itself in the first instance. It is nothing jointed; it flows. A 'river' or a 'stream' is the metaphor by which it is most naturally described. In talking of it hereafter, let us call it the stream of thought, of consciousness, or of subjective life."

The notion was immediately taken up into literature, as Morton Hunt writes, in a new stream-of-consciousness style by Marcel Proust, James Joyce, Virginia Woolf, and Gertrude Stein (who actually studied under James at Harvard).

10. ❧

Part of the sweep of new knowledge at the turn of the century, the *rinascimento* in memory studies, reached even the recluse Marcel Proust, a student of Dr. Henri Bergson. In his notebooks, Proust wrote of the new psychology of Bergson, James, and Ebbinghaus, and spoke of the mining of memory as an illuminating, important, and frustrating task:

A memory "presents itself so darkly at the bottom of consciousness . . . You can go on for two hours trying to repeat the first impression, the unseizable sign which says 'Deepen me' . . . It matters very little what the subject is. A hair, if it is out of reach for days, has more value than a complete theory of the world."

A pale, sensitive child, grown further inward as a young man, Proust was disappointed in love and in heterosexual and homosexual jousts. Following the death of both parents over a brief span of time, he gradually receded to his apartment, to lay abed all day. He rose only late in the afternoon if at all, and went out into Paris only at night.

Roger Shattuck once wrote that of the genuine classics produced in this century, Marcel Proust's *In Search of Lost Time* "is the most oceanic . . . a salvation by memory." The three-thou-sand-page work is not really a novel, but grew outside that defin-ition to what Shattuck refers to as a *roman à fleuve*, the book as river. And the spring from which it all flows is the memory—chilly, desperate, rich and made richer by reflection—of Marcel Proust.

His dark, Southern looks and soft manner, combined with a certain brilliancy in conversation, aided in his success as a slightly mysterious phantom of Paris society life. He looked haunted, and was reported by biographers to have been sickly from birth. Severe asthma limited his physical powers and sometimes crippled him. In addition, Shattuck wrote, "Mostly from his own testimony, we know that he was prone to hypochondria, voyeurism, and certain forms of sadomasochism. Psychoanalysts have produced resounding terms to apply to the roots of his condition . . . [but] I wonder if the technical vocabu-lary really improves on the term Proust's family applied to him very early and which he cites frequently in *Jean Santeuil: 'un en-fant nerveuse'*."

It is not entirely clear what set Proust off toward a deeper and deeper isolation. By 1897, when he was twenty-six, he had al-ready begun to rise in late afternoon and go to bed at dawn. By 1909, he told his friends that he would withdraw further from ordinary human life to carry out a work he had conceived and which could only be completed in fevered and solitary work. He lined the walls of his Paris apartment with cork to shut out noise.

Like the anguished Augustine, he was driven to renounce the world. "In the same way that blindness permits a greater con-centration on sounds, he was insulated from the distraction of new impressions," biographer Ronald Hayman writes. Proust dove into memory "as a means of reaching toward the mysteri-ous laws secreted behind surfaces." His few outings were likely to have ulterior motives: collecting new facts and tightening his grasp on memories of people he was using as models.

Proust finally sealed himself within the walls of his memory almost completely. Friendship, he said, "is a pretense, since, for whatever moral reasons he does it, the artist who gives up an hour of work for an hour of chatter with a friend knows he is sacrificing a reality for something which does not exist."

The watershed in all his ponderings about memory, and his intent to employ memory as the central element in his work, came as he wrote a piece of criticism and veered off into recollection:

"The other evening, having come in chilled, by the snow, and not being able to get warm, as I had started to read in my room under the lamp, my old cook offered to make tea, which I never drink. And chance had it that she brought me some slices of toast. I dipped the toast into the cup of tea and at the moment of putting the toast into my mouth and having the sensation against my palate of its sogginess permeated with the taste of tea, I felt a disturbance, scents of geraniums and orange trees, a feeling of extraordinary light, of happiness; I stayed motionless, fearing a single movement could interrupt what was happening in me, which I did not understand, but still concentrating on this taste of dunked bread, which seemed to produce such wonders, when suddenly the shaken partitions of my memory caved in, and it was the summers I spent in the country house I mentioned which burst into my consciousness, with their mornings, and drawing with them the procession, the nonstop charge of happy hours. Then I remembered: every day when I was dressed, I went down into the room of my grandfather, who had just awakened me and was drinking his tea. He used to dunk a rusk and give it to me. And when these summers were over, the taste of rusk dunked in tea was one of the hidey-holes where the dead hours—dead to the intellect—went to tuck themselves up."

From this recollection came, slightly altered to make the toast a more sensuous, seashell-shaped sweet, the scene which is the pivot of his verbal monument. His character relates that "one day in winter, on my return home, my mother, seeing that I was cold, offered me some tea, a thing I did not ordinarily take. I declined at first, and then, for no particular reason, changed my mind. She sent for one of those squat, plump little cakes called *petites madeleines*, which look as though they had been molded in the fluted valve of a scallop shell. And soon, mechanically, dispirited after a dreary day with the prospect of a depressing morrow, I raised to my lips a spoonful of the tea in which I had soaked a morsel of the cake. No sooner had the warm liquid mixed with the crumbs touched my palate than a shiver ran through me and I stopped, intent upon the extraordinary thing that was happening to me. An exquisite pleasure had invaded my senses, something isolated, detached, with no suggestion of its origin. And at once the vicissitudes of life had become indifferent to me, its disasters innocuous, its brevity illusory—this new sensation having had the effect, which love has, of filling me with a precious essence; or rather this essence was not in me, it was me. I had ceased now to feel mediocre, contingent, mortal. Whence could it have come to me, this all-powerful joy? I sensed that it was connected with the taste of the tea and the cake, but that it was infinitely transcended by these savors, could not, indeed, be of the same nature. Where did it come from? What did it mean? How could I seize and apprehend it?"

This is perhaps the most perfect description of the opposite of what Mr. M. experiences, of memory as utterly full. Proust did not believe that memory was a simple pictorial representation within, but understood it to be more:

"It is plain that the truth I am seeking lies not in the cup but in myself." The mind is the seeker and "at the same time the dark region through which it must go seeking and where all its equipment will avail it nothing. Seek? More than that: create. It is face to face with something which does not yet exist, which it alone can make actual, which it alone can bring into the light of day."

Part III

Part III

Tot circa unum caput tumultuantes deos.
So many gods, brawling around one poor man.
—Blaise Pascal, *Pensées*, 1660

11. ❧

The events at the change of centuries set the stage for the next act, and the characters of that time are still figures in the drama when we attempt to reconstruct the events of 1953 and the state of knowledge in psychiatry and medicine which might have led to the acts of the surgeon who operated on Mr. M.

William James's *Principles of Psychology* was still essential reading in psychology just before 1950, when Dr. William Scoville was a student. Freud was taught, and it was against Freud that Dr. Scoville and his colleagues in surgery reacted as they began to experiment with the excision of brain tissue. After James, Ebbinghaus, and Wundt, Ribot and Ramón y Cajal, scientific psychology somehow lost its way, and doctors entered the era of the lobotomy. The tools for further work were lacking, and the amassed theory appeared largely irrelevant to the dark prospects for the insane. Outside the academy and the laboratory there remained the asylum, where no hope entered.

It was this failure of theory which led physicians to begin their own ad hoc experiments, and between 1920 and 1960 they accumulated a history of their own clinical trials, ones which historian Elliot Valenstein has called their great and desperate cures. "Bizarre illnesses may require bizarre treatment," said Maurice Partridge in 1950, "and in psychiatry they often get it." Patients "show so often a stubbornness and resistiveness to treatment, they expose so clearly the ignorance of their pathology and etiology, that they arouse aggressive reactions in the baffled and frustrated therapist."

In asylums, the litany of aggressive "therapies" seemed like a prayer to torturers. Hard sprays of very cold water, shot from all sides, were employed till the patients became mute and appeared calmer. Electrical voltage was applied to various parts of the body. One psychiatrist noted that his patients seemed more placid when riding a jostling train, and so he began treatments consisting of shaking his patients for long periods of time. Another observed that mental patients who during illness suffered high fever seemed improved, and soon, according to Valenstein, "hot baths, hot air, radiothermy, diathermy, infrared-lightbulb cabinets, and special electric 'mummy bags' were all used to induce fever." Soon patients were given doses of malaria in order to induce the appropriate fever. In New Jersey, Dr. Henry A. Cotton determined that infection was the source of madness, and began to commit the insane to surgery to remove the unspecified outbreaks of bacteria. He removed teeth and tonsils, cervix and uterus, intestines and more. Finally, the most desperate cures began in 1935 when in Portugal Egas Moniz began the cutting of the brain as a treatment.

This kind of desperate medical practice did eventually swing around and cross, again, with legitimate scientific inquiry. Awaiting their moment were the surgeon Bill Scoville, brought up on lobotomies and the desperation of asylum medicine, and Dr. Brenda Milner, at an academy in Montreal where theoretical memory studies had become a center of interest. The nexus came in the person of the nameless amnesic, Mr. M. The study of memory was ripening in other ways, but the coming of Mr. M. represents a signal moment, the refutation of much science that went before, and the stimulation of much that was to come between the 1960s and the 1990s in memory research.

And so we return to Mr. M. It was, though the records appear to conflict on this point, probably on the early morning of August 25, 1953, that the young man came into his difficulty. That Mr. M. is unable to understand what happened, and what he has caused to happen, makes me want to tell his story, which would otherwise not exist outside the data of science.

He had lived in Hartford, Connecticut, or in the small towns beside it throughout his life since his birth in 1926. Hartford is a second-rank city, with a long history of standing in the shadow of Boston to its northeast. The neighborhood where he grew up was one of working-class bungalows inhabited by small-town New Englanders. The city lies quiet in the Connecticut River Valley, and was then a vision of homogeneous society when that seemed to be the ideal in America. The town was floated on insurance money. The tallest building in town, Mr. M. still remembers, was the landmark he would watch for as a child when they mounted the bridge to go into the city, the Traveler's Tower.

What thoughts Mr. M. himself has left under his pate about his life is only the cloudiest early memory, essentially all of it gelled before he was sixteen. When he now steps into the mirror of his own past, his foot lands on just this narrow spit of time. Sometimes he will speak of 1938 or 1942 in the present tense, then sometimes he will realize, if a little vaguely, that it must no longer be those years. It shifts in and out of focus as he speaks, but that is all he has, and so only images of forty-five and fifty years ago people his stories. We must construct our picture of Mr. M. before his surgery from what he can still tell us, and from inference, for his parents, grandparents, aunts, and uncles are all dead. There are no photos of him as a young man, and few in the years after his surgery.

Those who had known his father and young Mr. M. found the two alike in gesture and temperament: placid, short-spoken, and shy. They did not join in conversation easily, but could sometimes be started into stories of the family or some hunting trip. Mr. M.'s father, the son of a Louisiana French family, was named

Gustave, and in the 1930s he wandered up to New England with others in his family looking for work. He was an electrician, and not a bad one, as he seems to have made a living sufficient to buy a house and car and summer trips for the family. At least, he prospered in the years before the difficulties set in.

Mr. M. had been a strapping young man, according to the records: he stood about five feet ten inches tall, with a heavy, bony frame, a genial humor, and somewhat better than normal intelligence as we measure it by paper scales. Mr. M. has, and must always have had, an open smile and a gentle manner. He was never one for trouble. He didn't like "jive" music, but preferred Sousa marches. He stuck by the simple amusements of American village life of the time, such as the parade of popular radio shows: Arthur Godfrey, and "Major Bowes' Original Amateur Hour"; in the evening, "The Lone Ranger," detective Philo Vance, and "Adventure Tales." He was not cynical about these, even in the usually surly years between ages fifteen and twenty-five, but has said that he would work to untangle the plots night after night, guessing who was the perpetrator before the criminal was unmasked.

What few recollections anyone wrote down of his family's life in those years say that he lived first in the center of Hartford. He and his family were Roman Catholic and their parish church in Hartford, St. Peter's, was a pivot of their lives. His father was married in it, Mr. M. was baptized there, went to Mass there, had his first communion and confirmation there, and went to school in the building adjacent to the church. Young Mr. M. often played out the afternoons on the grounds of the church and school. It was supposed that young Mr. M. would also be an electrician, and he listened intently to his father's talk about the trade. He learned a few things and was proud to try to fix things as his father did.

After Hartford, the family moved just across the river to East Hartford, which was little more than a few buildings clustered on Main Street, and the village turned quickly to country. Houses and stores were interspersed with empty lots, then fields, then whole meadows as the road ran away from town.

In his small neighborhood in East Hartford, where he lived

during grade school, junior high, and part of high school, there were favorite places, such as Jean's soda shop, which was just across the street and down a block. He frequented her place daily, not only for the candy, sodas, and magazines, but because the proprietors, Jean and her husband, Bill, had traveled the world and had stories to tell. On Mr. M.'s side of the street, running behind the house, was a creek bounded by a narrow wood and a small bluff, which afforded sets for the games of soldier or hunter.

In summers, there were road trips in the Chevy coupe, into what passes for mountains in the East. He played cards, but not for money. Rummy was his game, at least in his younger days. His parents didn't swim, but they bundled him off early for lessons at a nearby public pool. He recalls with some relish swimming in a place called Glove Hollow. There, he said, he and his friends could climb the side of the dam and sneak over and slide down where the walls were sheets of water rolling down concrete slopes, to splash into the reservoir.

By his teenage years, Mr. M.'s family had moved away from the town to South Coventry, where they moved into an old country house on an unpaved road, and took their water from a well. He loved this place, he said, it had so many rooms, and his bedroom had a fireplace! And here he remembers the best feature of his youth, I suppose of all his life: the guns his father brought up from the south. He added to the collection himself later. There were pistols—.32, .38, and .44 caliber—and there were at least two rifles, both .22 caliber, one of them fitted with a scope. Behind the house, the field rose into woods and there was no road, so he could hunt there after a fashion. He saw birds, even pheasant, but couldn't hit them with pistols. He would have been unlikely to take them down with his small-bore rifles, either; he would have needed a shotgun, he said. But there were the squirrels he could chase. He recalled keeping his eyes open and listening for their "chirping noises" as they scuttered up and down overhead. He tried to guess where they were moving just ahead of their jumps, and he would guide his barrel in that direction, ready for a quick aim and shot.

There were trips he remembered from those years, both

hunting and fishing with his father and his own friends. Two of the trips in particular he recalls again and again, awake and asleep, one to a lake in New Hampshire, the other perhaps to a lake in Canada, it is not certain.

Among the coveted possessions of his childhood was also a pair of "shoe skates with fiber wheels," as he recalls with some faint smile even now. Nearby his favorite spot was the Skating Palace in Manchester, but he had saved stickers from all the places he'd been. His best sticker, he says, was the one from Miami, which he earned on a trip to visit his aunt.

Mr. M. had a girlfriend—he recalls her name! Mildred Carpenter—and even recalls the little drama of the time he met another girl—Here again is a name! Beverly McDonald—one day while out at the Skating Palace. Beverly, he learned, had her own skates and loved to go to the rink; Mildred, his current girl, hadn't and didn't. He sort of dropped the one girl for the other, he says now a bit shyly.

In his fifteenth year, he worked as an usher at the Star Theater, and says from that he learned two things. First, that they kept you moving so you didn't get to see the features. And second, that you took a big risk if you borrowed facts from the films to give answers in class. The facts of geography and history as depicted in film were likely as not wrong.

Mr. M.'s world in 1950s Connecticut included *An American in Paris* and the first IBM computer. Filter-tipped cigarettes were created, and *Playboy* magazine published in 1953 its first issue, with Marilyn Monroe as its centerfold. The mix was remarkable: Ralph Ellison published *Invisible Man*, while the movies of the year were *From Here to Eternity*, and *Mogambo* with Clark Gable and Ava Gardner. Jackson Pollock painted *Blue Poles* and Francis Bacon painted *The Cry of Pope Innocent X*. Sixty-four-year-old Charlie Chaplin was told by the Justice Department that he could not re-enter the U.S. from Europe unless he could somehow demonstrate to the Immigration Service that he was not a dangerous and unwholesome character.

Up to age sixteen, the only notable shadow over his life came when he was crossing the street one afternoon at age seven or so, when he was hit by a boy riding a bike downhill, too fast to stop.

He was clipped in the head, knocked unconscious, and the cuts on his face and head took seventeen stitches to close. (In the years to follow, all the neurologicial reports recall this accident like a talisman: it is suggested as a possible source of all that came later. Some scar tissue on the brain perhaps, that began to create erratic electrical patterns.) But little out of the ordinary otherwise. He expected to be an electrician and was certainly smart enough. He also thought of travel and horizons larger than the rim of New England.

On the morning of his sixteenth birthday, he recalls, he and his parents got into the gray coupe and headed for the city to celebrate his birthday. His father was driving as they approached Hartford from the north, from South Coventry, and came up to the Buckley Bridge. His mother was in the back seat, he in the front of the Chevy when, suddenly, his entire body seized up. His limbs stiffened and quaked, his head jerked violently back and forth. He pissed his pants and bit his tongue till it bled. His father swerved to the side of the road, surprised and angry. When the fit passed, they drove quickly to the hospital, which was in any case just ahead.

Until that moment, he had noticed only short moments of blankness, small interruptions. In the midst of conversation he would stare blankly for a moment, but was soon back, blinking, asking what was up. But from that day on, things changed. "What a birthday present!" he says, humor and dismay struggling together in his tone. He never rode in the front seat again. At school, the news came quickly enough, and while the teachers became cautious and over-protective, the other children said little to him, except for the occasional hurtful remarks. "You're going to wind up in the nuthouse, anyway!" they would shout.

He left high school, though he says it was not just because of the seizures, and did not return for three years until he went back to finish at a different school. When he did graduate, he was stung to learn that he would not be allowed to march up onto the stage to get his diploma. The teachers feared a fit, and so made him sit with his parents. At that time, too, he had to give up the thought of being an electrician. He couldn't work up on ladders for fear of a fall. The doctors who were treating him—it was the era of eugenic theories still—told him that he could never have sex and never marry because he might have more like himself.

From that time, his father began to have trouble. Mr. M. began to have minor seizures in which he blanked out for a moment several times a day, and major fits sometimes weekly and sometimes not for six weeks at a time. When Mr. M. had a seizure, his father headed for the neighborhood bar. This brought out nags and whines from his mother, which in turn sent him deeper into his bitterness. Mr. M. was the only child of the marriage and so, in just that awkward posture, the family froze. This isn't much to stand for his biography, but it is all that remains.

Deliverance was unlikely, but nevertheless, a surgeon at Hartford Hospital suggested a way out. He could try radical surgery, an operation never before tried in an area of the brain whose function was unknown. Perhaps it wouldn't work, but then, maybe it would.

At the time Mr. M. was growing up, the leading authority on epilepsy on this continent was Dr. Wilder Penfield, a Canadian neurosurgeon. He was perhaps the most eminent neurosurgeon of the day and was much admired by Dr. Scoville. (Penfield himself comes into Mr. M.'s story again later.) Penfield was carrying on brain surgeries that were not unlike those that Dr. Scoville was experimenting with in Hartford Hospital—the excision or destruction of certain small areas of the brain in an attempt to relieve epilepsy. Penfield wrote of epilepsy, perhaps ironically it now seems to me, that "the history of the multitude of therapeutic attempts to deal with epilepsy is not only a sad one, but a violent one. It is a history of maltreatment . . . largely a horrible account of persecution arising from ignorance, superstitition, and religious fanaticism."

Epilepsy has been one of the few disorders recognized since ancient times because of its stereotyped symptoms and its appearance in discrete bursts. Attitudes toward it remained not very different than they were in ancient days. In the code of Hammurabi there are laws regarding the marriage of epileptic persons and the validity of their court testimony. In the earliest sanitary laws of the Hebrews, epilepsy is mentioned as a familiar affliction. Hippocrates wrote the first scientific monograph on it in 400 B.C., "On the Sacred Disease," in which he stated plainly that it is a disease of the brain and not of the mind, and it has natural and not supernatural causes. As he described it, the disease appears as a convulsion of the whole body together with the impairment of leading functions. "The patient is dumb and loses consciousness, being insensible to sound, sight and pain. His body is drawn and twisted from all sides; in particular, his hands are cramped, his teeth clenched, he kicks with his legs and shows distortion of the eyes. Foam flows from his mouth, he suffocates and may pass excrements. In addition, coldness of the legs and hands . . . [and] turning upwards of the whites of the eyes . . ." He said there are sometimes auras which patients experience, by which "they know beforehand when they are about to be seized, and flee from men, either to their homes or to a deserted place, and cover themselves up." Lucretius just before the Christian era wrote of the sight of an epileptic who "wretches before our startled sight, struck as with lightning . . . He foams, he groans, he trembles and he faints. Now rigid, now convulsed."

Mr. M.'s father, Gustave, shrank from his son's disease. Mr. M.'s mother used to say that Gustave never recovered from the son's affliction, and he used to say that it was shameful to have "a mental" in the family, and an only son! He would be useless. Gustave did not want to discuss plans for their son's life, or his treatment, and turned all that over to his wife from 1943 onward. From that year also, things declined for Mr. M. The fits multiplied, and though young M. was able to hold jobs, they were limited by the risk of seizures. So instead of handling tools and guns as he ought, it was found that he could only throw rugs in the carpet section of a department store. He could wind copper in an electric motor shop. He could insert pieces on the assembly line at the Underwood Typewriter plant. By age twenty-six he had become a man of no great ambition, who lived at home within sharply narrowed horizons.

By the time I was able to inquire, I found that Mr. M. as a person hardly existed any longer, for he had become his illness.

The record, such as it is, is scientific and kept largely by Dr. Brenda Milner and Dr. Suzanne Corkin (Corkin has now spent a substantial part of her career tending to him, and testing him). It was they, who after some initial worry and refusal, agreed to allow me the freedom to read through what records there are of his life beyond the brittle pages of science journals.

Mr. M.'s seizures had, by the summer of 1953, increased to as many as ten minor blackouts per week, and in the worst months he was felled by a grand mal seizure once a week. This was not a great number as matters go in epilepsy, but they were a fearful disruption of life. His doctor, from neighboring Glastonbury, referred him to the neurological team at the Hartford Hospital for further treatment. The team there was headed by two doctors: William Scoville, whose chief specialties as a neurosurgeon were ruptured discs and lobotomies, and Benjamin B. Whitcomb, whose specialty was epilepsy. It was not Whitcomb, but Scoville that took the case.

12. ⤳

Family pictures show William Beecher Scoville as a thin, intense, and handsome man, almost unreal in his dashing appearance just standing by the hammocks at a summer retreat. Or there by the sleek car, with his wife, Emily, so beautiful beside him. He had come to Hartford, Connecticut, from the Philadelphia Scoville family, which had allied with the Beechers of Harriet Beecher Stowe and Henry Ward Beecher, the firm-spoken antislavery clan. An independent breed these were, willing to appear bold and dangerous. Dr. Scoville's father, Samuel Scoville, Jr., was something of a figure himself, being a naturalist, a lawyer, and a journalist who wrote a column called "First Aid Law" for a Philadelphia paper. Samuel was, besides, the well-known author of children's jungle tales that achieved some significant sales in the U.S.

Dr. Scoville and his new patient, H.M., moved within the same small piece of geography at that time, Hartford town, but at two different social altitudes. William Scoville had some wealth and social importance in Hartford, while his patient had none. But there was one ironic social link between them, unknown to either of them then or now. It was a mere thread that dangled from one stratum down into lives in the other: Dr. Scoville's wife, Emily, was related to the most prominent family of the region, the Cheneys, who owned the Cheney silk mill. Mr. M.'s mother was for a time one of the cooks in one of the Cheney households of the 1950s.

For his part, Bill Scoville was a wild fellow, if a respected inhabitant of Hartford. From the way he met his first wife—"he jumped on the running board of the car I was riding in," she said—to his pranks such as his climbing the cable tower of the George Washington Bridge at night and driving his red Jaguar at 135 miles per hour, the man seemed fearless. His special madness was sports cars. He had many cars, but more of Jaguars. The police knew him well, and set up roadblocks to stop his racing. He loved the machinery of the cars as well as the feel of the mechanic's tools in his hands. He had once told his father that he wanted to be an auto mechanic, but his father refused to allow it, and William went to medical school.

It was not unexpected that young Scoville might be an audacious neurosurgeon. He found that in surgery he might not have to discard his love for tools; there was cutting, coring, scraping, and sawing to be done. He invented some of the tools for his work and modified others. He also displayed his characteristic flash and charm in medicine, traveling widely, often to Europe and to South America and Asia as well. He was a founder of the World Neurology Society as well as its local counterpart. When he took in students at Hartford, he made a point to bring in young men from other countries.

Hartford Hospital was one of the nation's largest, and in the 1940s Scoville began there the first neurosurgery practice in Hartford. He and Ben Whitcomb, who was a brief-spoken Yankee and virtually opposite to Scoville in character, built the practice up together. The two doctors found each other difficult in many ways, but together they took in a varied and profitable trade. Scoville specialized at first in surgery on the spinal cord, particularly ruptured discs, for which he devised a new operation that had patients up and out of the hospital weeks faster than the conventional operation. Whitcomb became practiced in the surgery of epilepsy.

In their day, experimental brain surgery was most fashionable, and was spoken of with excitement. When they were in medical school, a neurosurgeon doing just this kind of work, Egas Moniz, had won the Nobel Prize. The award was for his creation of the frontal lobotomy, or leukotomy, which was his personal version of it. (That prize has in recent years been thought of as one of the most shameful and inappropriate the Academy has given, but at the time, it lent an air to the belief that mental phenomena could be controlled by surgery if only the right spots to cut could be found.)

Bill Scoville was strongly attracted to this sort of high-risk, potentially high-return work. Whitcomb did his own share of lobotomies, the partner admits, but nothing on the Scoville scale. The psychiatrists would call up, Whitcomb said, and ask them to pay a visit to the mental wards. "They would tell us they had a group ready. We did a lot of them in mental hospitals in Connecticut. One down in Middletown and one up in Putnam. We had to go in once a week. They had these cases lined up, and teams did them."

Whitcomb once wrote, at the request of the journal *Surgical Neurology*, an appreciation of his partner. Even in the prosaic pages of a scientific journal, Scoville's lawlessness was rather evident:

"A free spirit, unfettered by rules and regulations, Dr. William Beecher Scoville . . . has done much to influence the development of neurosurgery. He is an innovator, never willing to accept the status quo; and every surgical procedure offers a challenge to find a better method or to improve an accepted technique. Behind a facade of wild activity, driven by an insatiable ego, seeking better ways of doing things and espousing new ideas with their frequent and often angry confrontations," there nonetheless lies a nobility of spirit.

Scoville reported in scientific papers doing at least three hundred lobotomies, and probably he did many more. The work on schizophrenics and depressives over these years constituted one of the larger and more dangerous medical experiments in this century, in which tens of thousands, perhaps fifty thousand or more people between 1935 and 1960, were the victims of this blind surgery. The theory of lobotomy was confused at best, comprising conflicting guesses about how mental illness might be cured, if it ever turned out that lobotomies were actually successful. The evidence for their success, however, was based on quick personal assessments by the surgeons themselves.

Even at the time, it was the widespread belief of surgeons that the operation made one-third of patients much better, left one-third unchanged, and made one-third much worse. It was assumed that none of the patients would get better on their own, and it was believed the operation was progress of a kind. But patients were not systematically observed or tested, and much of the damage done was never recorded. We now realize that patients often do get better on their own, or at least the severity of their illness may improve and worsen by turns. In hindsight it is apparent, by their own math, that the surgeons did more harm that good.

"Well, thank God they don't do 'em any more. I am very glad the days of the lobotomy are over," said Ben Whitcomb in an interview at his home overlooking the Maine shore in 1993. "Those were terrible operations." There was one patient, a pianist who felt the surgery had succeeded, and she resumed her career in music. "For years, she used to send me Christmas cards," he said, still pleased.

"The worst one was the fella that was in a mental hospital and they called us down to do his lobotomy. The psychiatrists were the ones, of course, that selected the patients that should have it done. We did this guy, and he did very well. He was discharged from the mental hospital. He went back to Yale, and he graduated. Then one night this young man went out to the house of Dr. Thorn, who was the medical director of Yale University Hospital, and rang the doorbell. Dr. Thorn came to the door, the man shot him dead. The wife was screaming, she came in, and he shot her right through the head. Anyway, they picked him up in the railroad station half an hour later, and he had a list with him. Thorn was the doctor that sent him to the mental hospital, so he was first on the list. And I was the guy that did the lobotomy, so I was next on the list. . . . I hope he is still in the criminally insane ward. I told them if he ever gets out, I want to be notified."

By 1953, it was clear at least to the surgeons that the lobotomies had caused some severe problems. In even the best cases, the surgeries left the patients motiveless and confused at times, unable to plan simple tasks. But in this situation, Scoville saw a chance. Perhaps the seat of psychosis could be found elsewhere, outside the front of the brain. Perhaps, he wrote, surgery carefully done in other locations would not cause such distortions of personality. He knew in rough terms that the limbic part of the brain in the center of the head was the seat of emotion. It was the ancient brain core lying behind the frontal lobes, below the areas known to be important for perception and movement, and to the inside of the areas of speech. Surgeons had not attacked this area yet, and Scoville thought it worth a try. By the loose standards of the day, he counted his first efforts at these new lobotomies successful. Some psychotic patients appeared to improve after the surgery, though they were not carefully observed and were not followed for any length of time.

In a remarkable paper entitled "The Limbic Lobe in Man" published in 1953, Scoville wrote, "For the past four years in Hartford, we have been embarked on a study of the limbic area in man . . . We have isolated by the 'undercutting' technique, the anterior cingulate gyrus and the posterior orbital cortex in a series of fractional lobotomies performed on schizophrenic and neurotic patients." No discussion of ethics, or even passing comment, accompanied this striking statement.

He continued, "More recently, we have both stimulated and resected bilaterally various portions of the rhinencephalon in carrying out medial temporal lobectomies on schizophrenic patients and certain epileptic patients. I speak with all humility of the small bits of passing data we have accumulated in carrying out these operations on some 230 patients."

He concluded, "it is our belief, or perhaps I should say fantasy, that the limbic lobe of man may not yield up the secrets of smell, or of auditory hallucinations, or of fundamental mechanisms of mental disease and epilepsy. However, continuing limbic lobe studies may bring us one blind step nearer to the location of these deeper mechanisms. Who knows but that in future years neurosurgeons may apply direct selective shock therapy to the hypothalamus, thereby relegating psychoanalysis to that scientific limbo where it perhaps belongs? And who knows if neurosurgeons may even carry out selective rhinencephalic ablations in order to raise the threshold for all convulsions, and thus dispense with pharmaceutical anticonvulsants?" His operation, which cut out large sections at the center of the brain, "resulted in no marked physiologic or behavioral changes" he noted, "with the one exception."

This, the exception, is the first reference to Mr. M. in the scientific literature.

Ben Whitcomb had talked to Scoville about the case, and told him it was a significant risk, with no particular hope of benefit. Scoville went ahead, on an August morning, in an operating theater in downtown Hartford. Mr. M. lay awake, anesthetized only on his scalp. Dr. Scoville cut across his forehead and down the two sides, creating a flap of skin that he could pull down over the prostrate young man's eyes. This exposed an expanse of bone above the eye sockets.

Using a hand-cranked rotary drill ("which can be purchased for $1.00 from machine tool or auto supply stores," Scoville noted), he bored two holes over the eyes, each one-and-a-half-inches across. Into each hole he inserted a metal spatula. It slipped down in front of the brain, and was used as a lever to lift the front lobes of Mr. M.'s brain slightly. In computer scans done forty years later, the effects of this can still be seen: his frontal lobes are compressed upward and backward, making them appear smaller than they ought to be.

Having raised the front of the brain, Dr. Scoville cut out the fleshy lobes inside the brain, on either side, and then reached for the deep structures of the brain, among them the organ called the hippocampus—the sea horse—for its lovely curled shape. It sits upright in the middle of the brain; there are a pair actually, one on either side. The purpose of the hippocampus was not known at the time. Perhaps it had something to do with smell. It seemed specially sensitive to the touch of electrical rods and so might be the source of some epileptic seizures.

The first to record the appearance of this organ was Santiago Ramón y Cajal, the Spanish neuroanatomist of the last century. "I may say that one of the stimuli which led me to scrutinize the hippocampus and the dentate fascia," he wrote in his memoirs, "was the elegant architecture shown by the cells and layers of these centers . . . [They] are adorned by many features of the pure beauty of the cerebellar cortex. Their pyramidal cells, like the plants in a garden—as it were, a series of hyacinths—are lined up in hedges which describe graceful curves."

Dr. Scoville inserted a silver straw into Mr. M.'s brain and sucked out nearly the entire grayish-pink mass of the hippocampus and the regions leading up to it. On both sides. He drew out altogether a fist-sized piece of the center of the brain. In neuroanatomical terms, it was the area between the midlines of the two temporal lobes, and backward for eight to nine centimeters—thus removing most of the hippocampus, the parahippocampal gyrus, and the associated formations called the entorhinal and perirhinal cortexes, as well as the amygdala.

The bony young man on the table was awake through this, so say the surgeon's notes. And at what moment did memory slip away? Suddenly, as the lovely cells were sucked out? Or slowly, as fewer and fewer answers were raised from those regions? Did the feeling of power wane, the room close, and the ceiling drop? Was it a moment, as in the film *2001: A Space Odyssey,* in which Hal's memory cartridges were removed and he began to protest, but finally merely said, Oh! The answer is permanently dark, for the young man's awareness of that moment was flushed through the silver tube with his cells. In one sharp intake of air, he lost the world.

The abscission of these few organs (chiefly the curled hippocampus and the bundles of tissue leading to it) beneath the brilliant fire of the operating-room light marked both the loss and a moment of discovery in the small brick hospital in Hartford, Connecticut.

Here we should stop the story momentarily. If we may open this moment, surgically, to inspection, we will see several items of interest to those who would dwell on the mechanics of horror.

For example: there is virtually no other way that this particular large and bilateral lesion, a double-sided wound, could have happened by any other means than careful, skilled excision. The gallery of amnesics in science come chiefly from the population of people who have suffered severe brain injury from alcoholism, stroke, or catastrophic accidents, in which the soft brain was hammered against skull, first forward, then back with the recoil of the snapped neck. Or the case in which a fencing foil was run rapidly up the nose to puncture through the dura, the thick canvas covering the brain, or then a case in which a metal rod used in construction was driven, say, pneumatically, down through the brain case.

But you see how these wounds are all rather uncontrolled, grossly violating various discrete cortical structures. To reach into the middle of the brain, and excise just selected, well-outlined features of internal geography, and no more, was a remarkable accident! For it showed an appalling loss unlike any other before seen. It was one that could never be reproduced by injury but was so beautifully contained—no general madness, or agitation, or fits, or gross deformation of the personality followed. Instead a sudden, complete, and circumscribed loss of memory! It was thought by some boastful surgeons that this suction technique used by Scoville was clumsy. Walter Freeman, the world's greatest lobotomist, who became legend for carrying out his prefrontal lobotomies with a gold ice pick, once ridiculed the use of suction over the scalpel; he said it was like using a "vacuum cleaner over a bathtub of spaghetti," as Elliot Valenstein quotes him. Not always so, as we now see.

And here, another meticulous item. Dr. Scoville, upon finishing his interior work, left a little array of metal clips in the brain. They had no physical function, but were placed as markers just at the limits of his cuts, in case his method was a success. He could then show, just so, the depth of the cut.

At the end of the surgery, Mrs. M. must have retained some hopes, at first, before things began to go wrong. Some disorientation is to be expected, Mrs. M., and perhaps some temporary loss of memory. The mist may clear from the patient's eyes within a day or two. But a severe, full-body seizure overtook her son on the first day after surgery, suggesting that the seizures which the surgery was intended to end would continue. (In searches of the medical records since, there are a number of statements by Dr. Scoville suggesting that Mr. M. continued to have seizures, but fewer and less severe than before. It appears that the major seizures decreased in frequency from one per week to one every few months or longer. But, the seizures would no longer be the central issue for Mr. M.: he recognized no one. He knew not where he was, how he got there, or why.

His physical health was sound, and he had no gross physical deformities or anything that would suggest to the eye or ear that something was amiss. Dr. Scoville wrote of his recovery in the hospital, "his subsequent recovery was uneventful, apart from the grave memory loss." In the box on the hospital's form marked "Condition on Discharge," Dr. Scoville wrote "Improved."

In medicine, important things often remain unsaid. Doctors maintain a certain distance from their patients, perhaps out of necessity, perhaps from temperament. They look down at their papers, they speak in generalities. I suspect that a frustrating vagueness enveloped the case of Mr. M. His mother later wondered why she had agreed to the operation. She was angry at her husband for renouncing any part in the decision, leaving her with the whole burden of guilt. She went along with the suggestion of the surgeon, apparently, because she did not know what else to do. Dr. Scoville said there was a chance it would work. She stayed with the decision.

Mr. M.'s mind did soon become clear enough to recognize his mother and a few others. But most of the loss persisted, and extended through so many parts of his mental life that Dr. Scoville soon realized it was unlike anything he had ever seen. The doctors began to take notes, feeling for just how extensive the loss was. They stood as explorers at the the edge of an impact crater in the fog, trying to feel out the depth and radius of the great hole by hiking out along its edge.

The era in which Mr. M. lived, and the medicine of the time, betrayed him.

Of course, at the time there was no thought apparent about lawsuits or malpractice. That consciousness of doctors as potential renegades came later. In science, the story told about this surgery to wave after wave of students was cleaned up a bit. It was said that surgery for epilepsy in his case was successful but a serious side effect was found. True, but less than complete. There was a tragedy here with which the scientists who study the subject are uncomfortable.

From H.M.'s moment in surgery onward, every conversation for him was without predecessors, each face vague and new. Names no longer rose to the surface, neither histories nor endearing moments came any more. Reassurances of welcome had to be sought every moment from each look in every pair of eyes. The opening of that dark gap was for him cruel, but in science it was nevertheless the cracking of a door. His predicament began an era in the study of human memory, the successor to the era of William James and Wilhelm Wundt, the completion of their speculations, and answers to some of their first questions. Midwives to the event were desperation, pride, and foolishness, the hubris of myth. None of the people present understood the experiment they were conducting until it was over, and only then did they begin to learn the extent of the error.

When Dr. Scoville came home and told his wife of the surgery, she said that he told her in the form of a joke: Guess what, I tried to cut out the epilepsy of a patient, but took his memory instead! What a trade! She said he had no remorse, but he did soon realize how unusual the case was. He decided to acknowledge his error publicly and soon began to write scientific papers about the situation. We cannot leave the instant, though, without swinging the censer, shaking the holy water of absolution toward Dr. Scoville. He is dead and we must say, at least, that he did not hide his error. He called the eminent Dr. Wilder Penfield in Canada, who exploded into the telephone when he heard of the experiment. They resolved to follow the patient, to carry out extensive tests on this and all the other twenty-nine patients—all but H.M. were psychotic—who had been subjected to the experiment, to see if they had lost their memories as well.

("Testing" after lobotomies was so crude when done on severely schizophrenic patients that even an utter loss of the ability to form memories often went unnoticed. Of the other twenty-nine, only one patient, a forty-seven-year-old doctor who was schizophrenic, tested out clearly enough to say that he had a profound loss of memory from his operation, which was somewhat less severe than H.M.'s.)

In a paper Dr. Scoville acknowledged the surgery had been "frankly experimental"—an unusual admission in the cloaked and formal world of science. He pleaded in his papers to other surgeons not to do such experimental surgery anymore. The plea may have seemed unnecessary, but it wasn't. At least one group of surgeons reported similar operations after his.

Part IV

O! It comes over my memory
as doth the raven over the infected house,
boding to all . . .

                       —Shakespeare,
                       Othello

13. ⋘

I was again following the trail of Mr. M. in the mid-1980s, at Harvard on a fellowship with my wife, who was ill.

Some subjects cannot be related well in the abstract; memory is a subject whose personal power is too great, and our collective knowledge too small, to write convincingly only of third persons and theories. It rises from the ground of being like a spring newly opened and gurgling up, full of feeling. So rather than write about memory in the abstract I have told it here as personal episodes mixed with the material of the patients and scholars I visited. My own memories have been a part of the reason for exploring this subject, and through it I have found echoes of the experience of Mr. M. in my own life. Perhaps it would not be too burdensome to digress for a few more lines to walk through events which to me seem parallel in some respects to those which occurred to Mr. M.

I was unable to pursue the story of Mr. M. at that moment, as

I was distracted by my wife, Mary's, symptoms, which puzzled us, and by her worries about how they might affect our third child, with whom she was pregnant. An itching began to coat her body like an electric oil; it became so intense sometimes she could not put clothes on at all. She stood arms out, head back, and cried. She slept little at night, and in the morning was soaked with sweat. We went to eight doctors in sequence over a period of several months. Their remedies were a torturer's comedy of frustration: oatmeal baths and admonishments that women often overreact to pregnancy. Things spiraled downward, her energy flagged, though our third child came safely.

One February day, the doctor left for the operating room to do a biopsy on Mary, intended only to confirm a previous diagnosis. As the hours became heavy and grew dark I knew something was wrong. The surgeon returned to the chair in the narrow hollow hall of hell where I sat. There were problems, he said. A great mass. He spread his hands apart to show me. He cut out the mass, and he took her lung and other parts besides. In the following weeks it became clear that the surgeon had made a fearful error. He did not know Hodgkin's disease, but knew only years of surgeries on fatal lung disease. In a dreadful blank moment in the operating room, he forgot he was working with a disease treatable by chemotherapy but not surgery, different than the kinds of cancer he spent his entire career excising, lung cancers. When he took her lung, just the wrong move, it led later to seven more surgeries, and erased the possibility that she could ever have the strength or the breath to sustain the scalding and healing doses of radiation and chemistry. She was killed by the doctor on that day, but she walked, increasingly ghostlike, for three more years.

The darkness collected, but we planned a year of adventures. As we left for Cambridge, her weight had dropped from 120 to 89 pounds; we nevertheless began twelve months of music, of study, of children with shovels full of sand in the backyard box. In the basement of the art building, next to the William James house in Cambridge, my son used to run up and down the ramp, and I have since spent nights developing in dark trays the images I had taken of her, photographs of her at the end, weep-ing, cold, thin as bones. Because of this, I have, I think, understood the tale of Mr. M. better. I have felt it stirring something among my own memories.

In my notebooks from that Cambridge year, I see, references to Mr. M. are mixed in my hand and in Mary's. (We attended some lectures together, some separately, and the notes are interleaved. "The History of Life" was the slightly amused title Stephen Gould gave to his course at Harvard. From Mary's notes, I see, we heard that geology "discovered two great principles—ceaseless motion in the earth, and deep time, its ancient age.")

From those notebooks, I see we recorded a lecture by Suzanne Corkin on the brain and memory. "H.M. is a gold mine, the most famous neurological patient in the world." After one lecture, a student asked her "Is Henry still living in that nursing home in Connecticut?"

"Shh!" said Dr. Corkin as she glanced down the table at me. "There's a reporter present! We have to be careful." Thus I learned his first name is Henry.

Not long after, I flew to Montreal to visit Dr. Brenda Milner. It was the summer of 1989, a warm and clear July day. I met her in her office at the Neurological Institute. She is a Britisher transplanted to Montreal, and was the first to become deeply curious about Henry. She is a small, talkative woman with a droll sense of humor. Even when she is still, there is something about her expression and her rapid speech which make her seem as if she is moving, even at age seventy. Her face is rugged-looking, but tanned and rather youthful.

In 1951, she had already determined that she would begin studies of the mental effects of removing different parts of the brain, at the time an increasingly common procedure. Montreal Neurological Institute, where Dr. Milner worked, as did Dr. Wilder Penfield, the continent's most prominent practitioner of surgery to relieve epilepsy, was a terrific source of patients.

Dr. Milner's life is one completely absorbed in her subject—she has no children, no house, no car, no television. She had married Dr. Peter Milner, a physiological psychologist, but eventually they divorced. A frank and rambunctious woman, she told an interviewer of her chosen path: "I wasn't made to be a house-wife, babies don't appeal to me, and I don't consider children as innocent and charming creatures who grow up to become wicked adults." She was born in Manchester, England, in 1918 into a household breathing music; her father was a pianist and her mother a music teacher. She wasn't interested. She studied first mathematics, then psychology when she felt she wasn't making enough headway in her mathematical work at Cambridge University in England. She finished first in her class in Psychology.

Explaining her choice of study, she says that psychology was too theoretical and unlikely to be true: "I wanted to see what happens in a real brain."

She walks up the hill each morning to her work at the Institute and comes down again late in the evening. When Dr. Milner came to Montreal to finish her studies, she studied under one of the legends in the study of memory, Dr. Donald Hebb at McGill University. In 1949 he had offered a convincingly simple method by which we might understand cells, learning, and remembering. It is akin in the realm of psychology to the essential notion in biology that DNA is a long, coded molecule that can direct the activity of cells. Hebb said the chief mechanism of learning and memory is simply the strengthening of the connections, the synapses, between brain cells. Thus the repetition of a fact or experience will reanimate the same set of neurons and the links between them will get stronger, and thus more easily be recalled as a set.

Hebb was absorbed in the biological basis of mental activities and had got the surgeon Wilder Penfield to allow a psychology student to study his patients, each of whom had limited brain damage or epilepsy but was healthy otherwise, and were candidates for surgery in Penfield's huge round operating room at the Institute.

Penfield's most famous work was an extraordinary series of brain surgeries in that round theater, removing a variety of brain parts to dampen epilepsy. Along the way, he had accidentally encountered a phenomenon that intoxicated surgeons and psychologists universally. Brain surgery is conducted while patients lay awake, as there are no nerves to cause pain within the brain, and one day Penfield touched an electrified electrode to the surface of a patient's brain. To his astonishment, he heard his patient respond by retelling an event from childhood. Stimulation of the right side of the brain just above the ear of this woman he called J.V., Penfield wrote, "caused her to relive an episode of early childhood and to feel fear as she had felt it at the time of the original event. This was an experience that had reappeared to her in dreams also. . . . . Another patient, M.Bu., was caused to feel far away and another was reported who seemed to see herself as she was while giving birth to her child, in the surroundings of the original event." Unfortunately, these remarkable operating-table events turned out to be largely meaningless, because some may have been memories but many were an undecipherable mix of fact, fantasy, and memory that could not be caused again when the same spots were stimulated.

For student Milner, what she calls "Case Number One" came in 1952. A young man, an engineer, who now appears in the medical literature only as P.B., had suffered from epilepsy. He had a large piece of the left side of his brain, adjacent to the ear—a portion of the left temporal lobe connected to the hippocampus—removed. It was important that this surgery had taken only parts of the left temporal lobe, none of the right. Penfield realized that experimentation could be safer if only one side was taken, on the theory that if that particular part of the brain was more important than previously thought, at least one of the paired lobes would remain intact. The idea did not come from real knowledge that this was true, said Dr. Milner, "but it was a superstition that you could do without one side but not both." Of course, that did not take into account the possibility that the other side might have already been damaged but that the damage was unrecognized.

P.B. awakened from the surgery with a severe amnesia, the first time this had been recorded in the scores of such temporal lobe removals that Penfield had conducted. When Dr. Milner tested P.B., he asked her, "What have you people done to my memory?"

Case Number Two, referred to as F.C., was similar, and the patient underwent temporal lobe surgery on the right side in 1946 without notable side effects. But the surgery did not help the epilepsy, and a second operation in 1952 took the left hippocampus and nearby structures. He too came out of the surgery with a severe amnesia.

Dr. Milner reasoned that the hippocampus might be the common factor, and since both patients had full access to their memories that were laid down before surgery but had trouble forming new memories after, it was possible that the hippocampus's role might be to transform short-term memory into long-term memory. There was still much confusion about the finding, and it was unclear why patients with only one hippocampus removed would suffer amnesia. Similar operations in animals did not seem to cause such effects.

She sought confirmation of the idea, and the following year, 1953, Dr. Scoville called Dr. Penfield to let him know that he had done an operation in which he removed hippocampi and attendant structures from *both* sides at once, and the operation was a disaster. Penfield was angry, and interested. He passed the news to Dr. Milner and suggested she get on a train for Hartford.

"When I heard about Henry," she said, "I grabbed a few memory tests and hopped on the first train." She said she arrived at six the next morning and met him. His case, she found, was the purest loss of ability to form memories she had ever seen or heard of. "This was an intelligent, kind, amusing man," Dr. Milner told a Canadian magazine interviewer some years later. "But he couldn't acquire the slightest new piece of knowledge. He lives today chained to his past, in a sort of childlike world. You can say his personal history stopped with the operation."

She said that for Henry, every moment became like "that fraction of a second in the morning, when you are in a strange hotel room, before it all falls in place for you. It is very bad, because we became fond of Henry," she said. But because he repeated the same lines over and over and could remember none of those who worked with him, "we found ourselves beginning to regard him the way you would regard a pet. He lost his humanness. You can't build a friendship or any sort of human affection for the person."

She did say that in the early days, though Henry was still quite pliant and cheerful for the most part, he did sometimes make bitter remarks that showed his hurt, as when students mentioned "the case" in his presence. Henry would mutter, "I'm glad to know I've been an interesting case."

She says now with some emphasis, almost anger, that Dr. Scoville had performed a number of similar two-sided surgical removals but had done them in psychotic patients, so "he just didn't realize there was anything wrong with these patients."

She said, "This was a most radical procedure. He might have started with a more conservative operation, but not Bill Scoville. Henry was even worse than P.B." She recalls Henry's mother saying she had gotten cold feet before his operation and thought maybe he shouldn't have it. But she didn't insist, and it went ahead.

Dr. Milner noted in these patients—Henry, P.B., and F.C.—that their ability to form new memories was gone, but their thinking was intact. They had good short-term memory for such things as repeating numbers back, and recalling what has just happened. They also had not lost their personalities or sense of identity. But beyond a few minutes from the present, they became blank. "I began to wonder, what was spared in Henry?" This, as it turned out, was the crucial question.

Because Montreal was such a distance, it was difficult for the scientists there to carry on studies with him. Dr. Milner came down to Hartford again and brought Henry and his mother up to Canada for a week of further study. Milner has, since first meeting Henry and largely because of him, become one of the most eminent of the world's scientists of memory. But eventually the work shifted to M.I.T., which was less than two hours drive from Henry. First, the head of the Psychology Department, Dr. Hans-Lukas Teuber, took up the studies of Henry, and then, in 1966, Dr. Milner's graduate student, Suzanne Corkin, née Hammond, arrived at M.I.T. She had studied under Dr. Milner at McGill University in Montreal.

Eventually, the new headquarters for the study of Mr. M. became the Clinical Research Center at M.I.T., a small hospital ward which can care for patients at the same time they are under study at the university. It is located two blocks from the Charles River in Cambridge, amid the lively village that is the M.I.T. campus. The building is what might be called modern industrial—a clean, sand-colored building with no external ornaments at all, and glass doors opening on to a tiny, empty lobby with a few signs posted in proximity to an elevator.

It is here that Henry now travels once or twice or three times a year for his testing. These visits to M.I.T. are now the only salient feature of his life. Through the three decades of association with the university, M.I.T. has somehow entered Mr. M.'s world. Anytime Mr. M. is in surroundings that seem vaguely medical or academic, or when he is traveling anywhere by car, if he is asked where he is or where he is going, he gives a small smile, his eyebrows go up, and he guesses, "M.I.T.?" Mr. M. now has no family to speak of—his mother and father are dead, his aunts and uncles are dead, his cousins are dead or unknown, save one who never really knew him. So it is Suzanne Corkin at M.I.T. who takes care of Mr. M., works with the nurses at his nursing home, and handles matters of M.'s health and mood. She has become his family.

She is a short woman, and cheerful, though her students say she can be fierce. She has been chair of the Department of Brain and Cognitive Sciences at M.I.T., which she ran methodically. One piece of work that came with the job is the registry of brain-damaged veterans from three wars. There are more than four hundred of them with widely varying injuries. And there is Henry, professionally more valuable than any other work she has done. The question which absorbed her first was the extent of damage done by the removal of Henry's hippocampus and some central brain parts. Just what memory systems were damaged? Which were preserved? Could they be sorted into sensible categories?

"It became like a systematic parade through experimental methods, to try everything with him," she said. Through hour after hour, day after day, they tried to find some "explicit" memory that was spared, any memory of which Henry was conscious, such as events in his life, or facts, rather than his acquisition of new skills that he would never know he had. It turned out there wasn't any at all, "except for little fragments of things," said Dr. Corkin.

For example, Henry seems to remember something about a President Kennedy and an assassination, even though the killing occurred ten years after Henry's surgery. But on a closer reading of what Henry says when questioned about this at different times, you couldn't say he remembers the assassination of John Kennedy exactly, though he can sometimes answer questions correctly about it. If asked about an assassinated president he will remember McKinley, and recall an attempt on Franklin Roosevelt's life. If Dallas is mentioned, he may say, "Kennedy!" Asked for a first name, he may say Ted. Then, asked to reconstruct the Kennedy assassination, he will speak of an assassin who was trying to shoot the president but someone else got in the way; this is a description of the killing of Chicago mayor Anton Cermak, who was killed in 1933 by a bullet intended for Franklin Roosevelt—an attempt Mr. M. might remember from childhood. It seems as if there is some leakage from the world into his memory, some vague impressions which have become something like memories.

There are very few such fragments of "memory" that Henry has since his surgery. He knows what contact lenses are, and for a time when he was watching "Magnum P.I." on television every day, he could tell you that Magnum is the name of a detective.

It is not known how Henry could acquire these admittedly rare and vague memories. They may have been formed because some small pieces of his hippocampus and other central brain parts were left by the surgery, and still can function when bombarded with thousands of repetitions of a fact. Or perhaps the brain in some circumstances can shift the recording of memory to cells outside the hippocampus and its nearby structures, and use alternate routes.

At the same time, the general picture of his memory is that of total devastation. When asked by Dr. Corkin, who has worked with him since 1966, whether he remembers her, he

says, "You look familiar." From where? "Well, someone I went to
high school with, at East Hartford High." Dr. Corkin once asked
him how old he was, and he replied "around thirty." At the time,
he was more than fifty years old. She then showed him a mirror,
and asked if he still thought he was thirty. He said no, maybe
forty.

"We were trying to figure out whether Henry's memory impairment was really pervasive," Dr. Corkin said. Did it just have
to do with vision and hearing, or did it include loss of body
memory as well, such as loss of memory for touch or pain? The
tests were legion, and had fanciful nicknames such as "tactile
maze," "rotary pursuit," "famous faces," and "poison food
quiz." In the tactile maze, for example, Mr. M. sat at a table
with a black curtain near the edge of the table before him. He
put his hand through the curtain holding a stylus with a metal
tip, with which he had to follow a path through a maze on the
table, avoiding blind alleys. A bell rang when he went into a
blind alley; otherwise he had no way of knowing what was the
path and what was a cul-de-sac. Mr. M.'s time improved steadily.
He learned the skill, though he could not learn that the test existed or that he had taken it.

14. ✍

The overall impression of Henry to the time of the surgery was one of a mild man of some humor. This is still his way. His speech is clipped, something like what you might expect to hear from a New England farmer. Yep and Nope are often his only answers to questions.

Once as I was having lunch with him, I was a little put off by the dessert he had before him. "You like that stuff?" I asked him. As he spooned out the bottom and looked into the empty cup with a sly grin, "Guess so," he said. Apart from all else, I think of Henry's wryness. He is or was something of a bemused observer. By all accounts he always had his humor about him. And even now, despite his situation, he stands ready to amuse. Once, he and a researcher left the experiment room and the door locked behind them. The doctor paused for a moment and frowned. "I think I just locked my keys in the room," he said. "Well," said Henry, "at least you'll know where to find them." On another occasion, after the drive from Boston down to his nursing home, the researcher driving the car said, "That was a long trip. Are you stiff, Henry?" He smiled. "Nope, I haven't had a drop." The humor extends to his own condition. He is never without crossword puzzles, and has them by the book wherever he goes. Suzanne Corkin once remarked to him that he was "the puzzle king." "Yes," he said, "I'm puzzling."

Henry's deficit is most peculiar, at least to the uninitiated, because it left so much of his psyche still standing, while at the same time a yawning gap at the center in some way distorts even what seems normal, like trees around a crater bent back after a blast.

Henry could certainly still walk without difficulty, in his usual slow and easy gait. But after being in the hospital seventeen days, he still could not find his way to the bathroom and back. Shortly after he arrived home, his mother discovered that she couldn't let him walk two blocks down to the store, either, as he was unable to find his way back.

He could still read and write; his I.Q. was essentially unchanged, running about ten to twenty points above average. He had a subscription to a rifle magazine, and when he returned from the hospital, he still enjoyed reading it. But he would read the same issue over and over without knowing it. He used to mow the lawn, and he could still do it, never forgetting what needed cutting because the height of the grass made it apparent. But he could not remember when to cut it, or where the lawn mower was kept only moments after he put it there.

In later observations, it appeared that other patients had shown similar effects. Two patients uncovered by Dr. Milner in Montreal had surgical incisions limited to one side of their brains but, unknown to doctors, they also had destruction on the other side from some hidden natural injury. Their surgery took the only working parts of the middle brain left, and upon study they were found to have amnesic syndromes very like Henry's. One was a glove-cutter and the other a civil engineer. Both retained their higher-than-average I.Q.s and could do mental arithmetic without difficulty, and both returned to their work. The engineer, however, had to be demoted to draftsman. He could still draw blueprints but could not handle any jobs which had planning or administrative duties. Tasks were essentially limited to those in which concentration, aided by the cues available in the work itself, would allow him to work through to completion without guidance.

Henry's mother found that some unusual things happened when Henry was left to his own devices. She discovered, for example, that if he was at home while she was out, anyone who appeared at the door would be treated like a member of the family. Henry of course did not recognize any callers, neighbors or otherwise, and in order not to seem foolish he assumed that anyone calling for his mother must be a friend or relative whom he should have known. His natural desire to please led him to invite them in immediately, to sit and chat and wait for his mother's return. The strangers must have been rather puzzled.

Because he had hoped to be an electrician, he understood something of circuits and did not shy away from mathematics. But he now was forced to put his arithmetic to rather pitiful uses. When asked the year, or his age, he now tried to calculate it from the clues around him. He might glance at the date on a newspaper or infer dates from other things that were said.

But really, Henry has no notion of what year it is, or how old he might be. Once, Dr. Corkin showed him some old family pictures that she had obtained before his last relatives died. In a short session in which she asked whether he recognized anyone in the snapshots and when they were taken, he guessed at years in which they might have been taken—and said at one point 1936, forty years before. At another point he guessed in the sixties, or the seventies. He has even guessed they might have been shot "in the mid-eighties, 1985 or '86." The year in which the interview took place was 1977.

When he saw a picture of himself next to his aging mother, he recognized her, but not himself. The man next to her, he said, must be his father.

Age and year guessing are perennial favorites with nurses, doctors, and others who visit him. It is a simple, somewhat amusing demonstration. Once when he was fifty years old, a student had the following completely typical exchange with him:

"What is your birth date?"
"Well, February 26, 1926."
"And how old are you now?"
"Well, I don't know."
"How old do you feel?"

"I don't remember the year now . . . I think I'm about thirty-three."

"So you feel like you are thirty-three?"

"I feel like I am, but it's sort of a natural deduction."

"From what?"

"Well, I feel I am older than twenty-nine . . . and it's not thirty-six. . . ."

"How old were you when you had your operation?"

"Gee, I don't know."

"If you were to tell the date that sounds most likely to you that you had your operation, what date would you choose?"

"I think of '78, right off."

The year of that interview was 1974. The president had just resigned and the Watergate crimes and their consequences were dominating the news, but he did not know anything of them, and guessed that the word Watergate must have to do with an engineering project to hold back water. In the same interview, however, he recited the Gettysburg Address, named the year that Lincoln took office, and described the contents of the poem "The Midnight Ride of Paul Revere." He can call up some knowledge from the time before his surgery, though most is forgotten. That which does respond to the call is entirely time-bound: he was once asked about the music coming from a nearby radio, hard rock-and-roll; he referred to it as "swing." When guessing years, he most often picks dates in the 1930s. Henry's room at home, which Dr. Corkin visited in the 1970s, was like a museum of teenage life in the 1940s. According to her notes of the time, his bedroom at home, before his mother died and before he went to live in an institution, did not "seem to be the bedroom of a forty-six-year-old man, but rather of a boy in high school."

She wrote, "The room itself is small and tidy, and gives the impression of being very static. Almost everything in the room is old. The only new evidence I can see that this was 1972 rather than 1952 was a Remington electric shaver." The shaver, she noted, he apparently did not know how to use. There was among the debris an old crystal radio set, a large tin racing car from the early 1940s, and a model of a plane from the early 1950s.

For Henry, no such independent life is possible. He is utterly dependent upon others, on the nurses who care for him, on Suzanne Corkin. He can do nothing without direction, worrying always what "went just before" and "what I should be doing next."

On one summer morning when Henry was home with his mother—she was almost ninety years old—she collapsed in the kitchen. She fell to the floor and did not move. Henry came in and sat by her, but did not know what to do. When someone arrived later and found the two of them on the floor, Henry was asked why he didn't do anything. He said he thought she was just resting; more likely he was simply immobilized.

His inability to move himself from square to square is among the chief sources of trouble for him. Without instructions, Henry would be innocent of motive.

He lived with his parents from the time of his misbegotten surgery to 1980, when at age ninety-five his mother was too old and too ill to care for him anymore. Henry moved in with a woman, Lillian Herrick, a retired psychiatric nurse and a distant relative of his.

Until that time, each morning, Henry would wake up and ask, "What am I supposed to do now?" His mother often said it was a burden, worse than having a child because even children have some sense of what comes next.

Mrs. Herrick described her way of dealing with it, convinced as she was that some day he would remember. When he asked, she would say, "Well, you tell me." Henry would not know and would answer, "Well, I'm having a debate with myself." Mrs. Herrick would counter, "Now, Henry, don't debate because you have the answer." Henry would then begin to guess. If it was morning, and he knew it, he might rub his face as if to say he needed a shave. "Well, that's one thing," Mrs. Herrick would say. Henry would try again. "Oh, you want me to sit down to breakfast!" And she, who had proved Henry did know after all, would say, "There you have it, Henry!"

The same routine would ensue, no matter the subject. Whatever Henry could not infer from conversations around him or from expectant looks, he had to learn by enduring the tedious back-and-forth.

Henry's father died in 1966. But he could not remember it. This upset his mother and Lillian. When his mother also became ill and went to the hospital, Henry sometimes made the mistake of asking about his mother or father.

> *Lillian:* Well, now you tell me, Henry.
> *Henry:* I'm having a little debate with myself.
> *Lillian:* Don't debate, Henry. You know your father's gone and your mother is in the hospital.

Summing all his minuses, and adding other fragments of knowledge gained in the meanwhile, it is possible to draw a coherent picture of Henry's forgetting. When his mind was cored, what was excised was the hippocampus and its attendant parts, and we can now see their central role.

It is not the place where long-term memories are stored—he still has some few of those from before the surgery. It is not the place where the programs controlling walking, speech, gestures, and such other mechanical memories are held—he still has these. It is not the instantaneous memory which can hold on to the things that are in the current spotlight of attention—a few seconds' or minutes' worth, until something distracts us. It is not even the place where memories of who he is are stored, because he does know some things about himself: he knows of his loss, even if his understanding is vague. Each of these is a separate engine of memory, and must be located elsewhere than in the hippocampus. Though Henry's life is now lived in this two-dimensional sheet of the present, there are still things about him that are whole. His personality is intact, in a way, along with his physical frame. Though few have suspected it, personality turns out to be only a mode, a manner, and not the content, of being.

This is one of the lessons of the story of Henry, that the brain has many separate engines, and memory, too, is not one power, but a battery of different powers. Dr. Milner tells the story of her surprise one day when Henry appeared to retain a memory for perhaps fifteen minutes running. She told him the number 584 and asked him to try to remember it. She left the room and returned a few minutes later, and Henry quickly blurted out the number, 584! Dr. Milner found that if Henry worked steadily to keep the number before his attention continuously—like batting up a beach ball repeatedly to keep it from hitting the ground—he could remember it. He still has what might be called "immediate memory," and using it, he can hold some small fragments for some minutes. The moment his attention is broken, however, the ball drops and is gone.

What was at the center of his mind and is now missing for Henry is the engine of memory which we use to catch the events of the world as they go by. As the stream of life moves rapidly through the net of our senses, we snare some of it, hold it so that it may become gradually etched as a permanent part of our mental collections. It means that no events passing before Henry can ever be retained: no persons, no sense of where he has been and what has happened even in his sharply narrowed space of life, none of it may do more than appear briefly then be lost into the dark.

The understanding that one organ in the brain would do all this, and that there are several other engines of memory besides this, was new to scholarship, the gift of Henry to science.

Henry stayed home for a time after his surgery, but in the 1960s and 1970s, he was sent during the days to a retarded children's home. He was able to do some work: inserting metal clasps into folders at the rate of 450 per day; he could separate white and colored paper from carbons at 100 pounds per day. He needed to be reminded from time to time what the task was. When he visited the toilet, he came out and could not find his work station even though it was marked with a flag. Until his job was described to him again and the materials were put in front of him, he did not know what to do.

He was surrounded, say the notes of the staff, by "feeble-minded girls in adolescence" who were "already obviously promiscuous." But Henry was always a gentleman to them and unable to get into trouble even when provoked. When asked about sexuality, he has over the years repeated the same formula, that the doctors told him he shouldn't or couldn't. The nursing staff never noted any evidence of sexual activity on his part, alone or in company. His sexuality may have been excised along with his memory.

Dr. Hans-Lukas Teuber, aware of the gray days which comprise Henry's life, once thought that perhaps especially emotional or vivid moments might somehow make an impression on Henry. Perhaps strong interest and attention, with a chorus of emotion, could spark whatever small pieces of memory hardware were left in Henry's cranium.

Driving Henry up to M.I.T. from his nursing home one afternoon, he got an unexpected chance to test the idea. It had been raining heavily nearly all day, Dr. Teuber said, and at one time the sky became so heavy with water that his vision was totally obscured, and a second later cleared only for him to see a car in front of him veer out of control, spin completely around, and shoot across in front of them up the bank on the right side of the road. The car turned over on its side and its tires blew out with a noisy blast. Dr. Teuber hit his brakes and skidded to the side, avoiding the overturned car in front of him with some difficulty. He then jumped from the car to give aid to the mother and daughter driving the damaged car. The women screamed and were shaken, but not seriously hurt. The police came, the car was righted, and eventually Dr. Teuber and Henry got back on the road. The doctor was soaked to the skin, both shirt and trousers. The excited conversation continued for some time as they drove. But eventually after twenty minutes or so, the topic shifted and there was a lull in the conversation. Dr. Teuber decided to test Henry. "Why am I all wet, Henry?" Henry quickly said, "Because you got out," then he paused briefly and added, "to inquire about the way."

Later that night Henry remembered the incident no better. Dr. Teuber had found Henry sitting in the TV room, and asked if he was watching the tube. Henry was holding a towel and face cloth. No, he said, the nurse had told him to wash up for the night, but he could not find his way back to his room. It was immediately across the hall.

A few days later, Dr. Teuber again wandered over to visit Henry at night and found him sitting "rather forlorn in his room at dusk, near his puzzle, but not working on it. Asked if he was feeling okay, Henry said he had no physical discomfort."

What, then?

"Well, mentally, I am uncomfortable. To be so much trouble to everybody. Not to remember, and I keep debating with myself if I said anything that I shouldn't have, or done something I shouldn't have."

Dr. Teuber reassured him, and said they could call Henry's mother later that evening.

Henry thought for a moment, and said he was now having a debate with himself. "About my dad. I'm not easy in my mind, on the one side I think he has been called. He's gone. But on the other hand," and Dr. Teuber said he noticed Henry trembling, "I think he's alive!"

His father had been dead for four years. Perhaps the dread which crept into Henry's thoughts came from Dr. Teuber's inquiry: he had mentioned only the possibility of calling Henry's mother. His father had not come up, and Henry quickly guessed at what might be wrong.

15. ✑

There are others who have severe amnesia, though perhaps not as severe as Henry's. Each has some similar blankness and some different presence. I have met several and read about others.

In 1987, as I began to explore the nature of memory and its absence in amnesia, I visited Nick A., another amnesic patient who had been found in the 1960s, and had been studied at M.I.T. Known in the scientific literature as N.A., his home is in San Diego, where he lives in a trailer park.

Dr. Corkin, who knew Nick, said the feel of Henry's room when he was at home—that of time frozen—was much like that of Nick's. As I entered Nick's small white trailer, I was confronted at once by that feeling. Here, Nick's past was spread out upon every surface, as if the contents of his memory were spilled out into the room instead of held within his head. The table, the television, the floor were all covered with small objects: shark's teeth, model airplanes, statuettes, colored rocks, a tarantula in glass, and a mounted wallaby among the spreading carpet of miscellany.

There was one room in this trailer that could not be entered at all, because it was filled a foot deep in touchstones from his life. I recall Nick standing at the door, pointing to objects and trying mostly without success to recall the story of each. As he talked, his face flushed beneath his blond military-style haircut. He struggled for seconds, then for minutes, and still could recall little. Embarrassment spread like a stain over the conversation.

After a tour of the poignant debris, Larry Squire, the San Diego scientist of memory who had driven me out to visit Nick, said of him, "For Nick, every topic is memory." His memory loss is not so bad as Henry's, but is still debilitating. After prompting he recalls one incident that troubles him still. A few years ago he had just bought a bicycle, and then rode it up to the fast-food restaurant near his house. He went in, sat down, and looked out the window: a teenager was toying with his new bike. But Nick didn't react. As he watched idly through the window, a group of boys gathered around the bike. "They fooled around with it for a while," Nick said, "then all of a sudden one of them got on the bicycle and rode off." He continued to watch without special interest. "I'd forgotten that it was my bike. It just didn't dawn on me that it was my bike he was taking."

After I had talked with Nick for an hour and a half, we were ready to leave, and Dr. Squire asked him a few questions. "When we first came in, we introduced you to him, and talked a little bit," said Squire, pointing to me.

"I don't remember his name at all," Nick said.

"Where is he from? Do you remember? What is he doing?"

"Is he writing a book, or a newspaper article?"

"What's the subject?"

"I don't remember."

To carry on each day, Nick must chain oft-remembered items together, doing them in order and concentrating on using each task as a cue to the next. Washing up leads to bed-making which leads to laundry. Notes don't help. He can't remember that he has them in his pocket, and if he does find them, he is uncertain as to what they refer to, or whether they are current or old.

Dr. Corkin says emphatically, however, that Nick is not amnesic in the purest sense. He can and does form many memories. At least, Nick can live alone, and if he finds himself at a loss can remember to call his mother.

One of the most unusual books on amnesia was *The Man with a Shattered World*, published in 1972, by the famous Russian psychologist Alexander Luria. Written after thirty years of working with a single patient, a man wounded in the Second World War whose memory was destroyed, it is one of a pair of such books Luria wrote about extraordinary cases. "Neurological novels," Oliver Sacks called them, though they are nonfiction: "Neither Freud nor anyone else has given us a case history of this length. . . . No one had conceived a neurological novel before Luria." Luria was both a physician and psychological naturalist. But as one doctor said of him, unlike the naturalist who studies species, Luria was concerned not with species, but "with a single organism, the human subject, striving to preserve its identity under adverse circumstances."

The tale in the little book by Luria is that of a young man, Sublieutenant Zazetsky, who was shot in the head while crossing the frozen river Vorya to attack German troops on the other side. He ran forward, head down, and the bullet broke through the top of his skull, just behind the center on the left.

The injury was in a location quite different from Henry's. Many of the effects were quite different, but Zazetsky also had a catastrophic loss of memory. "In the first days after the injury he could understand nothing and remember nothing. Glimmers came back to him out of the darkness over months and years. He said, 'I'm in kind of a fog all the time, like a heavy half-sleep. My memory's a blank. I can't think of a single word. All that flashes through my mind are some images, hazy visions that suddenly appear and just as suddenly disappear, giving way to fresh images. But I simply can't understand or remember what these mean. Whatever I do remember is scattered, broken into disconnected bits and pieces.'"

In words which seem hauntingly like things Henry says, Zazetsky said, "Again and again, I tell people I've become a totally different person since my injury, that I was killed March 2, 1943, but because of some vital power of my organism, I miraculously remained alive. Still, even though I seem to be alive the burden of this head wound gives me no peace. I always feel as if I'm living out a dream—a hideous, fiendish nightmare—that I'm not a man but a shadow, some creature that's fit for nothing."

As he speaks of his condition, he realizes that it must not be a dream. "My therapist tells me we have been at war for three years and I've become ill ... So that means I haven't been dreaming all this time. Of course not. A dream can't last this long or be so monotonous. That means I've actually been experiencing this all these years. How horrible this illness is! I still can't get a grip on myself, can't figure out what I was like before, what's happened to me."

His sight was damaged, and he lost all vision on the right side. Because the feeling and intellectual sense for his right side lay in the destroyed part of his left hemisphere, he could no longer see or even imagine his right side. He lost his sense of location of parts of his body and what their functions were; for example, when he needed to defecate he sometimes could not remember what it was that he needed, the fact that he had an anus, or where it was located.

Still, the front of his brain was whole, so he was generally aware of his situation and kept planning work to overcome his "illness." With the effort of years he found he could read sentences, but only very slowly, one letter at a time. If he thought of it, he could not write at all, but if he simply held the pencil to paper and thought of the letters, his hand could write them seemingly on its own, though so slowly that it might take a day to make a page.

Over the course of twenty-five years, he managed to write down scraps of reflection that cover three thousand pages. Dr. Luria edited it for him, supplying his own commentary and interlacing it with Zazetsky's text.

Luria was one of the more prolific and original of this century's psychologists. He wrote of long personal relationships with patients, of the history of Russian psychology, of the structure of the brain; he was a man who conveyed at all times an intimacy with this most intimate of organs.

In the tale of Zazetsky, he described the soldier's injuries by laying out how damage to various parts of the brain would produce different effects on understanding and behavior.

The visual area of the brain, at the back of the cranial dome where scenes are first analyzed, is "a remarkable laboratory that breaks images of the external world into millions of constituent parts," writes Luria. "It had been left undamaged by this patient's injury."

Adjacent to the primary visual area at the brain's back is the area where the second wave of processing occurs, forming features into objects. "If one applies an electric shock to the primary visual cortex (this can be done during a brain operation and is absolutely painless)," Luria writes, "glowing points, circles and fiery tips appear before the person's eyes. If, however, one applies the shock to any part of the secondary visual cortex, a person sees complex patterns, or at times, complete objects—trees swaying, a squirrel leaping, a friend approaching and waving."

Thus injury to the primary visual cortex essentially erases vision, but damage to the secondary area produces a peculiar effect. Persons and objects are still seen, but the area which melds them from parts into complete images is damaged. This was not Zazetsky's damage, but Luria described it for comparison: "A person's vision undergoes a bewildering change: he still distinguishes individual parts of objects but no longer can synthesize them into complete images; and, like a scholar trying to decipher some Assyrian cuneiform, can only surmise the total from the separate parts."

There is a third, still more comprehensive level of analysis within the brain. Here was the damage to Zazetsky. In this area are combined and summarized the conclusions of the areas upstream of it—vision, touch, hearing, and so on. This third tier of organization was the last part to develop in evolution, and is not mature in humans until sometime between the ages of three and seven. They are called "zones of convergence" or "association areas."

It is precisely this area, as Dr. Luria writes of his poor patient, "that the bullet fragment destroyed in this patient's brain." His vision, apart from some deterioration on the right side, still worked. He continued to perceive discrete objects because his secondary visual cells were spared. But after the injury, he could no longer combine his impressions into a coherent whole. He was aware of his body, but could not tell his right from his left. If he were to try to sort left from right, as he did when his doctors pressed him, he had to first try to locate his arms in terms of a system of spatial coordinates, starting from some fixed point, and work assiduously to lay in each object and note its relation in the frame. He would quickly lose track.

In daily life, he could not put on a robe by first looking at it, because he couldn't determine which arm was which. He could not tell time from a wall clock because, while he could see the numbers, he could not decide which way the clock was turning.

He could not fathom simple relations expressed in sentences. We are accustomed to the balletic turns of phrase: daughter and mother, mother's daughter, daughter's mother. But Zazetsky couldn't do it. He knew there were mother and daughter in the phrase, but could not decipher their relation. Is an elephant bigger than a fly, or smaller? A fly bigger, the same size, or smaller than an elephant? Lieutenant Zazetsky could not tell you. "I'd think and think about these but get all confused. My mind seemed to be galloping back and forth so fast my head ached even more. So one way or another I'd make mistakes, and I still don't understand these things . . . All I could figure out is that a fly is small, an elephant is big, but I don't understand bigger and smaller." The relations between things were lost, though the things themselves were not. For him they floated, curiously independent of one another, things in themselves but unable to be manipulated jointly to compare them.

"A person with such an injury," Luria writes, "finds his inner world fragmented; he cannot think of a particular word he needs to express an idea; he finds complex grammatical relationships unbelievably difficult; he forgets how to add or use any of the skills he learned in school. Whatever knowledge he once had is broken down into discrete, unrelated bits of information. On the surface his life may appear no different but it has changed radically; owing to an injury to a small part of his brain, his world has become an endless series of mazes."

Henry M. is lost down a somewhat different set of corridors. His injury was not to the dome of his brain at all, but to the midbrain below the dome, where events are bound together to become lasting memories.

Unlike Zazetsky, he can see well enough, and decipher objects, and know their purpose. He can do sums, and can determine whether a fly is bigger than an elephant. The world is not a puzzle of parts which must be forced together; this perception comes as naturally and inexplicably as it does in you and me. When we are used to having these powers under our hands, we have great difficulty in imagining ourselves without them. Henry too has these powers; for him the scene of the moment is full and rich. But it is thin as vapor. His loss is a more pure loss of memory. When he turns, the whole scene before him is gone. Whoever and whatever is inside his zero-thickness world is present only as a white marble museum figure. Each has no history, and he can never know them but by their appearances. He cannot go backward to another time when he was touched by her or him, he cannot recall a kindness or a cruelty. There are scattered items left in his recall from before the surgery in 1953, because they were laid there by his hippocampus before it was taken. They are obscured somewhat, and some bits that he might have had are lost, because he did not recall them often enough to maintain any access to them. For example, Henry did know Dr. Scoville before the surgery; there are within Henry's grasp pictures of this man. But it is only the man before the act, the confident surgeon who was going to try something to banish the fits.

"Scoville?" Henry smiles with complete incomprehension. "Oh yes, I know him."

Did he see him, did Scoville help him after the surgery?

"I don't know. Did he? I guess he must have."

Beyond the injuries of Zazetsky and Henry, there are others who have some distinct combination of deficits and abilities. For example, injury to the front of the brain, the "prefrontal cortex" as researchers redundantly refer to it, produces other distinct lapses. Such injuries were commonly and intentionally inflicted on mental patients and other uncooperatives under the name lobotomy or leukotomy. In the most severe cases, perception and memory remain relatively normal, while the patient becomes unable to form intentions, project plans into the future, or exercise conscious control over his own behavior.

In my researches, I traveled to Toronto to speak to another amnesic, Kent C., a patient of the eminent Canadian psychologist Endel Tulving. Dr. Tulving, a thin, graying, and goateed gentleman, is an immigrant who escaped the Soviet Union in 1945, and went to Canada to begin his career. Now at the University of Toronto, he is perhaps the world's senior researcher on memory; at any rate, he has that classical look of a European professor together with a light, slightly amused accent and attitude.

In 1984, he learned of a young man, a black-leather-jacket sort of fellow who had run his huge motorcycle off a curve three years before. Receiving massive head injuries, he was unconscious for hours. When he was called back to consciousness, he knew nothing. Only gradually did fragments of memory return.

When Tulving heard of this unusual patient, he went to visit him. Kent is a young man of average height, brown hair, and a babyish face that is dominated by very large brown eyes. The eyes seem blank, neither searching nor evaluating his environs as most people's do. Rather, he appears not to see or hear most of what goes on around him. He does of course both see and hear, but he waits without response, just as Henry seems to. He sits quietly, as if that were the safe thing to do rather than move out into the uncertain waters of a flowing, little-understood conversation. As he sits, he smiles a little, and seems expectant.

By the time Dr. Tulving met him, Kent's days as a hell-raiser—riding big cycles, drinking, and smoking marijuana—were over. Tulving asked about his life, but Kent knew nothing of his own prior image. More than a decade before, Kent had gone to college, yet as they sat in the small hospital room, he couldn't recall much of it:

"What happened after Sheridan College?"

[Pause.]

"Do you remember anything that has happened between going to Sheridan and being here today?"

[Pause.] "No."

"What have you been doing since then?"

[Kent laughed.]

"You find it difficult to answer this question. Why? Can

you tell me why it's difficult?"

"My memory."

"Can you describe your memory? What does it feel like when you're trying to think about the time between Sheridan College and being here?"

"It's just bare."

"What do you mean, bare? Can you give me some sort of comparison, an analogy?"

"It's like being asleep."

"But at the same time it's not like being asleep, because you're here right now, talking to me."

"That's right."

"And you have a very clear idea of what's happening right now?"

"Yes."

"If, say, a Martian comes in and asks you, 'Kent, what are you doing right now,' where . . . can you describe the room?"

"Well, I'd say in a twelve-foot-square room, I guess. It's got one window in it. And there's a professor in here with me. I don't remember his name. And he's talking to me, about things I am supposed to remember." [Kent laughed.] "And I can't remember."

"How far can you reach back into the past?"

"That's a good question, because I don't think I can remember even five minutes ago."

"Can you remember coming in here . . . something I did at the time when I turned on the tape recorder?"

"No."

"Now let's go into the future. After we finish here what will you be doing?"

"Probably eating my lunch."

"You have not had lunch yet today?"

"I don't think so."

"Then, after lunch, what will you do?"

[Pause.] "I don't know."

"When you think about tomorrow, what will you do?"

[Pause.]

"What will you do tomorrow?"

[No answer.]

"Well, can you try to describe the state of your mind, how it appears to you?"

"Blank . . . It's a big blankness sort of thing."

"What is it similar to?"

"Like being in a room with nothing there, and having a guy tell you to go find a chair, and there's none there . . . or it's like swimming in the middle of a lake. There's nothing there to hold you up or do anything with."

"Can you see the shore?"

"Yeah, but it's way far away . . . Too far away."

Henry himself described this state in a similar way; it is like perpetually waking from a dream, he said, and he always feels suspended in that just-waked uncertainty before he knows just what day it is, or what he must do.

It is a sort of perpetual uncertainty that is difficult to understand until you first confront the state in the flesh. I was able to do this some seven years after I first heard of the man and his state.

He raised and threw back his head,
like a silkworm searching for a leaf;
reflected for a moment, to recall more clearly
to his memory a fact of which he retained only
a shadowy idea; remembered the circumstance,
had a vague and momentary recollection of
the person; passed on to something else,
and thought no more about it.

—Alessandro Manzoni, I *Promessi Sposi*

16.

When I first met him, Henry of course was much older than the images I had created for him in my researches. He is now a pale bear of a man with missing teeth. Friendly eyes peer out from the slack, yellowish skin of his face. He still has an openness which speaks of geniality. His eyes are wide, he smiles easily.

My first memory of the man himself is from the end of a corridor—a ward hallway that I have seen in scores of hospitals and clinics, in climates warm and cold on four continents—whose many plain doors open into a hall the color of which barely registers on the eye, much less in memory. (Is there some mad architect loose, climbing catlike into the offices of other architects at night to transmute their plans into his perfect rendering of nondescript?)

From the corridor's end, I could see him framed in the empty rectangle of gray; there was nothing else in the hall but his figure, sitting in a small chair and bent over a book. He seemed motionless for the entire time it took me to reach him. His still silhouette grew from a small, stooped black-and-white outline to the full, fleshy figure of a bear, seated.

He wore brown trousers and a checked shirt, glasses with thick rectangular frames of tortoise-shell brown. His pepper-gray hair rode far back on his forehead. His mouth was puckered, and in a moment I saw why—his front teeth, top and bottom, were missing.

His head was down as I came near. Was he reading? For it seemed to me then that a man so without mental mooring would have trouble doing even that. Could he remember how to read, and then what he read as well? I imagined him dropping the front of a sentence by the time he picked up the back, in a sort of Chaplinesque sketch of a man with packages.

This, before I learned of just what he could do, of what powers a man may have and still get no useful grip in this world. Even with many faculties intact, he can live only if protected from the rigors of the world at large, as an infant in some way, except that he knows he is no infant and no full man either.

Despite all the secrecy, there on a placard to the right of his hospital room door was written plainly his full name. Henry M——. I was a little surprised to see it, and Dr. Corkin noticed me looking at it; she was dismayed. Upon arriving at his chair, a schoolroom model with an arm and quarter-table on one side to write on, I found him totally absorbed in playing at crosswords.

As I later realized, these have the thankful property in Henry's world of being self-cueing. There before him are all the answers he has labored over and won, in neat rows crosswise and shoulder to shoulder with the empty frames which tell him, first, that the puzzle is not yet done, and second, which clues to study. Ah, if the world were a crossword!

He is a soft, doughy man. About five feet ten and a little round, though not fat. The most lasting impression of him is of pale eyes and brows lifted, an expectant look. He waits like a dog for attention to fall on him, for a word assuring him that he is doing fine. More, that he has not done anything wrong. He worries about this.

He looked up blankly. First for him is a question: Have I come to see him, or am I just passing? The feet coming toward the chair signal their intent in advance as they aim toward him. He raised his head just before impact.

From the first look, Henry can tell whether he is supposed to know the intruder. Those who know him act that way subtly, a slightly lifted countenance, eyes that search for his. Thus Henry knows that he should know. My approach caught Henry's attention only when I was close, and this moment for Henry is critical in any encounter. In these few seconds many things can be determined, and for the rest, his train of guesses can be pulled out and the work of figuring out what these new people want begun. Henry enjoys encounters for the most part; they are puzzles providing whatever uncertain pleasures he may derive from other people. And though in theory he should be unable to remember his situation in life, I believe he takes some pride in the constant testing of him, again and again, that he says "can help me, and help others, too," as he put it so often. It is a conversational talisman, a mantra to attune himself to his fate.

"Hi, Henry. Do you know me, Henry?"

"Well, you look familiar—" He pauses artfully, hoping his questioner will supply the necessary information.

"You want to go for a little walk, Henry?"

"Well. All right." He slowly nods yes, but seems a little uncertain.

"We want to do some word tests. Do you think that might be fun? I know you like word games."

"Yes, and I do crosswords, too." You see, he has already forgotten. Now he notices his crossword book. He reaches for the book and smiles; it pleases him to be able to contribute to a conversation and refer to an accomplishment of his own. He does not realize how many times a day he repeats this same sentence—"I do crosswords, too"—thus erasing in the minds of his interlocutors any credit he might gain for his ability to converse normally.

Slowly, Henry stands, lifting himself first from the chair by his hands, then reaching for the support of his walker.

"We're going to go to the testing rooms, Henry. Do you know the way?"

Henry pauses, working to fetch the information. "No, I don't."

But he has been here scores of times and Henry's body knows. Or more exactly, the part of his memory devoted to retaining skills and procedures, which is quite separate from his memory for facts or episodes of life, does know the way.

"Well, which way do you think?"

As he repeats, "I don't know," his body nonetheless turns right, and his foot slides subtly in that direction.

"But you do know, Henry. See, you're already headed that way."

From the time I first saw him the question which has dogged my thought of him has been: What is it like? Can I imagine anything that would resemble his state and what it would appear from the inside?

I believe his interior view is still seamless. He does not seem to ask about what is not there before him very much, he has only what he has. All of what approaches him in the world is a surprise, and perhaps partly because of that, he has no motive force of his own. He doesn't know where to start and instead waits passively for someone's attention to fall on him.

When it does, he tells each and every visitor who talks with him a series of anecdotes. It may be that these are what shuffle through his mind over and over, throughout the days. He repeats the stories which have now become Homeric in their rote form, nearly word by word each time, as if recalling the history of the tribe, the song he cannot forget.

He will tell that his father came from the South. His grandfather was a sheriff in Thibodaux, Louisiana. He was shot while collecting a prisoner from Texas. The bullet lodged close to his heart, too close to remove it, but he lived. His mother was born in Manchester, Connecticut. (This is at least partly true: the censuses from LaFourche Parish, Louisiana, show that his father was born in Thibodaux, Louisiana, in 1892. His grandfather was born there in 1833 and was a deputy sheriff.)

He tells of living in South Coventry, Connecticut, where he could go out behind the house to shoot his guns. He had a rifle, "One with a scope!" he says, always enthusiastic at this point in his story. "And I had handguns, too! A .38, and a .32." And on. He may have a few dozen little fact-stories. They do not have the feel of real memories any longer. They are blanched and brittle. Again and again, he turns to them.

Poor, familiar sticks! I do not know if he runs his hands over them again and again when he is quiet. He may have once; now they appear as others talk, share reminiscences, and laugh at each other's stories. This is all he has, so he flashes his light on them again and again. Twenty times a day? Ceaselessly over the past forty years.

The words to him still do not seem to have lost their sense, but to others they now are like songs, little tunes he hums when eyes turn on him. Henry, along with others who have eclipses that approach his in totality, are caught just there, on the cusp of the present, awkwardly balanced in the rapid stream of time which they cannot catch nor stop.

Once, he surprised the scientists with a passing insight to his state. He spoke of the fear that tugs at his ear constantly: "Right now, I'm wondering, have I done or said anything amiss? You see, at this moment everything looks clear to me, but what happened just before?" His soft voice sometimes seems plaintive. "That's what worries me." And he spoke of the boundaries of his world. "It's like waking from a dream; I just don't remember. . . . Every day is alone, in itself. Whatever enjoyment I've had, and whatever sorrow I've had."

Now in midday, Henry sits with his crosswords and sometimes watches television, though he does not remember the shows. He simply turns on the set and whatever channel is on, he will watch until interrupted. As Henry wakes each morning, the world comes back to him, but not the meaning. What am I to do today? He does not know. If asked where he is, he cannot answer. Who is taking care of him this week or this month? Or who has made his bed, cleaned his small bathroom, set out his clothes in the morning? If his clothes were not laid out, would he know to get up and put them on?

Henry does not really recall as we do when queried, simply turning mentally to pick up an answer without effort. For him answering questions is desperate work: he does not know what people want of him; he tries to please, but he just does not know, as he puts it, "what went just before." We mistake this: we are certain our answers are correct in some way, but he has no such assurance. Which of the things he is saying are inventions? Which real shards of a broken memory?

I have asked him quite directly what is on his mind, what he is thinking of, but it is difficult to ask the right question. He looks at me expectantly, alertly trying to work out what I want. He may say he is thinking of his crosswords, or of the nurse then walking by and whether she may be going to the kitchen. But after a thought or two about the genuinely mundane, he soon veers back into one of his old rutted roads. The old stories always seem to satisfy people, or at least people tire and turn away. I do not know what is on Henry's mind—no one knows—and he has trouble telling about it except in brief phrases which then ride like boats on the swells of his regular stories.

What might Henry have been without his amnesia? Would it have been a pleasure to be in his company? Could he have done something remarkable in his life or given an assist to someone else? Amnesia is a book left open to an early page, a letter unanswered, a telephone off its cradle and a voice calling down the wire.

Where we have lush forests of recall to turn to, he has a few gray, dried trees. We move through time as through a landscape, able to see things coming from far away and able to orient ourselves with expectations of what will come next. For him, the world does not come naturally from a landscape, but bursts out of a wall of mist. Each moment is a surprise, a new puzzle to be worked out from a quick glance at the paltry evidence at hand as it comes rapidly upon him. Still, astonishing as it may seem, Henry can report back to us at times from this precarious ledge. There is one particular exchange in the record which I find extraordinary.

Once in an interview he wandered away from the topic he was supposed to be working on, like a horse wandering from the track to the grass. He told British researcher William Marslen-Wilson that he worried about getting things wrong in testing, or whenever anyone asked him a question. And was not all of life now a continuous questioning? He said he worried about what had happened to him since his surgery.

"What happened to you? Why do you think this is?" asked researcher Marslen-Wilson.

"Well," said Henry, "I think of an operation. I have an argument with myself right there—Did the knife slip a little? Or was it a thing that's naturally caused when you have this kind of operation?"

"That caused what?" Marslen-Wilson asked.

"The loss of memory, but not of reality," Henry said.

He is in reality, he told Marslen-Wilson, but without the memory you have to think things out more to understand all of what's going on. Even then, when you "get all the ends and put them together, then think about them and decide," you may still get it wrong.

"Most people think," Henry said, beginning a little lesson in philosophy of mind, "most people just think things through once; they're able to pick it out from memory and everything they have." But he must struggle through all the clues—from questions asked, from the way people look, from all the marks of the text, like a mind reader watching all the cues that are too common for their dupes to notice.

It is a constant effort, Henry said. You must always "wonder, how is it going to affect others? Is that the way to do it? Is it the right way?"

Marslen-Wilson, sounding lost, asked, "What has to be done?"

"Whatever it is that you're sent to do," said Henry. He is a man who has done little but fetch answers and follow instructions for forty years. "Any job that you're doing." Asked if he worried about these things a lot, struggled with his thoughts to get right answers, he said yes, all the time.

"But why?"

"I don't know," said Henry.

Memory—the hearing of deaf action,
the seeing of the blind . . .

—Plutarch

17. ❧

After I had met Henry for the first time, I began to wonder about what his experience of the world is like. I tried to imagine myself as Henry and exclude what I could not know if I could not form memories. It is a partial consciousness which is difficult to imagine. But for one element of Henry's mentality, one researcher tried to imagine just how Henry might estimate time's passage.

"Without the normal recall for events, how fast does time pass for H.M.?" asked Dr. Whitman Richards of M.I.T. in a paper in the 1970s. "Does one hour, one day, or one year seem just as long to this unique individual as to us?"

Dr. Richards asked Henry, in a standard test, to estimate time. He said go, then waited for an interval, and said stop. Then Henry was asked to give go and stop signals that were spaced equally to the ones that he had heard.

Henry, it turns out, has a normal sense of time for twenty seconds. He can recall the task and reproduce the lengths of time very well. But after twenty seconds—essentially the period of immediate memory that does not depend on the hippocampus—his sense of time moves into a different mode. Past twenty seconds, Richards wrote, assuming that any real sense of Henry's time estimation can be worked out, "then one hour to us is like three minutes to H.M.; one day is like fifteen minutes; and one year is equivalent to three hours for H.M." In this way, it may seem to Henry that about five days have passed since his surgery in 1953.

I thought about Henry's interior experience, and began to seek out the experience of loss which others have felt. How would Henry's loss compare or contrast with that of someone who lost sight, or hearing? One extraordinary book which describes the experience of blindness parallels Henry's experience in some ways. John M. Hull writes in *Touching the Rock* that "My perception of time has undergone a change since I lost my sight. . . . Perhaps all severe disabilities lead to a decrease in space and an increase in time. I think of my friend Curtis with his multiple sclerosis. Without his mobility machine, his range is about twenty yards . . . Time, on the other hand, has strangely expanded. It takes him forty-five minutes to tie up his shoelaces in the morning. It doesn't matter. He does not get impatient. He just does it. That is how long it takes to tie shoelaces." He himself felt the same inflation—tasks take as long as they take, which is much longer than for sighted people, but "You are no longer fighting against the clock, but against the task. You no longer think of the time it takes. You only think of what you have to do . . . Time, against which you previously fought, becomes simply the stream of consciousness within which you act."

Things must be similar for Henry. Time includes only now and the twenty seconds just past. For Henry, time has not expanded, but has been reduced to almost nothing, so that even the smallest task fills all the time that there is. So long as the cues are in front of him in the form of a crossword puzzle with blanks, or an experimenter with cards, or a hallway with distance before him, he will move along as if there was no time, only a task. In this way, we can imagine Henry not disturbed moment to moment about his state. The boundaries of time for Henry are full, not empty. We may even imagine him happy for the most part. But I found that this is not always the case—Henry does have sudden outbreaks of emotion.

There comes with these narrowings of time a dependence John Hull found depressing: "The world which remains, then, is one's own body, the introspective consciousness. . . . Blindness takes away one's territorial rights. One loses territory. The span of attention, of knowledge, retracts so that one lives in a little world. . . . Only the area which can be touched with the body or tapped with the stick becomes a space in which one can live. The rest is unknown."

Henry's blindness is to the passage of time, and his world has retracted to a similar tiny world, one of his own limited consciousness, or alternately, one in which he tells stories over and over to make a small social space for himself in time with them. There comes with the narrowing of time a dependence. Time is shared; it is as social as a pool filled with bathers. But Henry cannot join.

Blind people gradually lose their memory of faces to put with people they hear, Hull says. They become disembodied voices. While Henry can see the faces, and hear the voices, it is the persons—their history and intent—which have disappeared for him.

I suspect that loss of sight, however damaging, robs less of a person's standing in the world than does the loss of time. Those who are stripped of sight or hearing remain full characters, active and full of thought. Blindness and deafness may erect a barrier between a person and the world, but both person and world remain. Helen Keller broke through that screen.

For a person lacking memory, however, while there is no barrier of sense, the world evaporates like steam. Henry may recite facts. Where is Toronto? Who is president? The machinery of thinking may stand. Long division is still possible, and by some tricks of a different kind of mathematics, emotions can be aroused—anger and love. But all escape as fast as they are created. They are useless.

Because of this, I think that perhaps time is the human element, one not given to other animals in the same way, and the one in which rationality and relationships are embedded. Thus memory may be the founding trait of humanness.

In excavating memory, in the digs of archaeology and the digs of poetry, I find strong images through history. Memory buried, when unearthed, frees the ghosts which have been incarcerated inside the earth: the sleeping bodies of Pompeii, the dignified kings wrapped in cloth and stone in Egypt, or the dark and leathery preservations which rise out of the bogs of Ireland and northern Europe.

The poet of memory in our time, and perhaps more than that, is Seamus Heaney, tousle-haired farmer's son, who now teaches at Harvard and Oxford. He must be the only poet in this millennium to write a poem about the ancient method of memory taught by Simonides in Greece and Cicero in Rome: "Memory as a building, or a city, well lighted, well laid out, appointed with *tableaux vivants* and costumed effigies . . ."

Of the Celts, buried in the bogland, he says:

> I step through origins
> like a dog turning
> its memories of wilderness
> on the kitchen mat:
> the bog floor shakes . . .
> water cheeps and lisps
> as I walk down
> rushes and heather.

Long patches of his work about the land in Ireland and northern European could be read as a litany of metaphors for memory:

> Each open pool
> the unstopped mouth
> of an urn . . .
> Earth-pantry, bone-vault . . .
> Sword-swallower, casket, midden, floe of history . . .
> outback of my mind.

Whatever images and metaphors there are for memory, I keep coming back to the water; it is pure medium, a transformer and shape-changer. It can break hard sea-nights over its knee and make of them gentle patterns—ropes of light slipping across the black silk surface. In the daytime, water takes its color from the sky, the tanned sandy bottom, and the distributed green particles suspended within it. Leaf green, sky blue, stalk gray, sand yellow, rock brown, and all riffles of white around its neck. Seeing into memory, or into time, is like looking into water. There are places so clear you can see red coral heads a hundred feet below.

The sight includes you—marrying the depths of water below you, the heights of sky over you, and the breadth of earth's rim around you. It is a map in which you live. But here amnesics, as Oliver Sacks writes, are lost mariners, steeped in fog while we play around them in the deep light.

"The lough waters can petrify wood," Seamus Heaney writes, "old oars and posts over the years harden their grain, incarcerate ghosts of sap and season. The shallows lap and give and take," he wrote of the memory held in Irish ground, held in the acid pores of the bogs. "Butter sunk under more than a hundred years was recovered salty and white, the ground itself is kind black butter . . . the wet center is bottomless."

(My wife, Mary, and I read these works to each other during our 1984 season at Harvard, as I worked on memory. We studied in Heaney's class and took him to dinner for Portuguese squid. And from that time are some of my most vivid memories of the last time with my wife. I remember coming home late after dinner one evening with Heaney. Because my wife had several surgeries, she finally had a great wound in her chest. Her lung was gone, and there was a cavity in her chest open to the air. It was a hole that extended inward through the cavern where the lung used to be. Daily I had to pack this empty core at bedtime with rolls of saltwater-dipped bandages, pressing them far inside her chest with forceps. On that night, the ritual was a little uneven, as I was drunk and the saltwater spilled down her side; she laughed and we went to the refrigerator for more wine. We sat on the blue couch looking out to the black trees moving in the wind. Our youngest, one-and-half-year-old Sean, lay on the chair. Wet with wine, Mary said that she had thought of what she would miss most. There were many faces reflected in the wine-fogged window as she enumerated them. But finally, she turned her head to Sean. "Most, I will miss seeing him grow." I photographed her face many times, and her shrunken body. Together we marked on film the picture of cancer coming down on her. Then there were nights I spent extracting her images in the darkroom, swishing pictures and pulling them up from the chemical baths, like relics from the bogs.)

In the records of Henry there are dreams recorded. At night he is freed from his body and glides through woods, and over lakes, and by the ocean shore. But here too, as in his waking hours, things are not right. Mr. M. can sleep, but of what might he dream?

In the night, there is no refuge: his dreams as he reports them are as spare as his waking, oft-repeated scenes. It is like the dreaming of young children before they gain the power of memory and narrative. His dreams are striking confirmation of the theory that dreams are part of the process of consolidation and rehearsal of memory, which is apparently required to keep it for long periods of time. In Henry's dreams we find only that which he would still have to rehearse—images from his life as a teenager and young man.

A few hours after Henry has fallen asleep, he is shaken awake by researchers plumbing his dreams, and his ghosts appear. Here are the records from several nights:

> "Henry, Henry, Henry!"
> "Oh!"
> "Were you dreaming?"
> "Yeah."
> "What were you dreaming about? . . ."
> "I was having an argument with myself . . ."
> "About what?"
> "What I could have been . . . I dreamed of Pennsylvania. I dreamed of being a doctor. A brain surgeon. And it was all quick. Flashlike, being successful. And living down that way . . . tall straight trees. . . ."

Same night, 3:52 A.M..

> "Henry . . . What's the last thing you remember?

"Woods."

"What do you remember about the woods?"

". . . Just coastline. Differences of coastline of Maine, one being flat coastline and the other being all very hilly, stony coastline."

"Is that all you remember?"

"'Bout all, maybe the question mark of which has been, or why one's one, and not the other . . . . to be at ease with yourself with it."

Another night, 3:00 A.M. at M.I.T.:

"I dreamed of a lake and a boat."

"Go on."

"Fishing in it, but swimming in it, too. Diving off the boat into the lake. And it was away from the shoreline . . . it was clear water . . . don't go too far, the boat might get washed away, and you couldn't get to it.

"If you dive deep enough in a certain area . . . you can go through a tunnel and come up inside of a cave . . . you have to swim underwater to get to it.

"I am always wondering how the air inside the cave is freshened . . . You could dive under and . . . you wouldn't know where you were going. You could be going into an area where, if you did come up, the air was stagnant or something, and you were breathing it and it wasn't the kind of air you should breathe. Or you got trapped and you couldn't swim back."

At 5:50 A.M.:

"Henry? What is the last thing you remember?"

"You remember after you've forgotten. No matter what you forget, you always remember . . ."

"What is the last thing you saw?"

"Well, I can't think of anything . . . bed . . . going into it . . . I'll let you know . . ."

"What?"

[Here, Henry has begun to play a game with his ques-

tioner, which apparently the questioner never under-
stands.] Henry said, "I said, I'll let you know the day after
yesterday."
"Huh?"
"When's that?" Henry asked, trying out his joke.
"I don't know."
"It was the day after yesterday."
"I don't know."
"Today."
"What was today?"
"The day after yesterday."

Another night, 5:15 A.M.:

"Henry . . . were you dreaming? Were you dreaming,
Henry?"
"I think so, yeah . . . about what I wanted. What I could
have been. You know, I sort of, uh, one part about it in the
dream I thought of Pennsylvania. . . .
"I thought of being a doctor . . . A surgeon. A brain surgeon."

At 1:21 A.M.:

"Henry, were you dreaming? Were you dreaming, Henry?"
"Uh huh."
"Can you tell me about it? . . ."
"I'm afraid."
"You're afraid? What are you afraid of?"
"The wind was blowing hard."
"Where were you?"
"Big wind, big waves. . . ."
"Where?"
"The lake."
"Look around you. What do you see?"
"Woods."
"Were you doing anything?"
"Walking around . . ."

"You said you were afraid of something . . . of what?"

"The trees falling down."

"Because of what?"

"The wind."

His dream sessions reveal, again and again, his fear of "what went before." He has often spoken to researchers about the general doubt he lives in. He knows that everyone else can reach back into memory and know the answers to questions, the solutions to situations, the next thing they are supposed to do. They can, as Henry might say, "find the right way" easily. But in his mind "answers" do not appear from memory. Rather, choices, possibilities, appear from his thinking brain, his forebrain. Perhaps his most common expression of all—with Henry this means many repetitions per day—is "I'm having a little argument with myself." He knows of possible answers to questions, of possible reasons why things are, but he can only guess among them. This is frustrating to him, and behind this guessing when asked questions or presented with situations lies a worse fear— that Henry may have done something bad, and he will not know it. Still worse, that people will not tell him. It can be seen in the following sad, painful exchange:

At 4:30 A.M.:

"Henry. What's the last thing you remember, Henry?"

"Well, I know I wasn't asleep just now . . ."

"You weren't?"

"When you said '4:30 . . .'"

"OK."

"Because the other gentleman came in and I believe that I was going to tell him the same thing that I told you, about the canal . . ."

"Oh, you were?"

"And I didn't. I couldn't remember it."

"You couldn't remember what?"

"What I said about it . . . and repeat it exactly the same way. I don't know if I swore or something like that. But—

something like that. And I couldn't tell anymore; I told you both to shut up."

"You told us that?"

"I don't know. I think so."

"I see."

"I'm not sure. 'Cause I could have said shut up or something like that . . ."

"But do you remember what you were going to tell us about?"

"Well, I told you, I think I told you about building a canal in Canada . . . to open up the western part . . ."

"Henry, are you sure you told me this before?"

"I think I told you this before, and you went to get the other gentleman in, and told me to tell him. But I couldn't repeat myself then."

"How long ago was this?"

"I can't exactly remember the time. It was tonight I know. I said I couldn't repeat it—and I believe—I don't know if I swore or what, but I told you 'shut up' or something like that."

"I see."

"I couldn't repeat it, and I think I swore . . . told you both to shut up and get out of here . . ."

"You've been trying to remember it all night?"

". . . And put two and two together and try to place it and put it all together, in a way."

"Oh really? . . . OK."

"Sorry that I couldn't [remember] it."

"All right, Henry, that's fine."

"Sorry, though. That's the way I am."

"Don't worry about it, Henry."

"It's easy to say that. I still have a little confusion there—I should have remembered it in a way, but I couldn't."

"But Henry! You did remember!"

"I did?"

"Sure. I mean, how could you tell me unless you remembered it? You did a good job, Henry . . ."

"I did? See, I think that I didn't."

"Well, I can let you listen to the tape sometime. OK?"

"Well, I'm sorry—and the way I'm possibly acting now, because I don't remember."

"Henry. Don't worry about it. You're doing fine. I mean you are doing very well."

"Oh, I warned you!"

"You don't have to warn me. You're doing very well."

"Well, I just wonder."

"I'll let you go back to sleep now."

"Because I imagine the disappointment that can cause—I said it once to you, but I didn't repeat it over. And I believe I swore. . . ."

"You can rest at ease. You didn't swear at us."

"Well, when you both came in. . . ."

"Well, when you're asleep, sometimes you think something happened that actually didn't. So don't worry about that."

"Well, I wonder though. That's it . . . I'm sorry though."

"OK. We'll let you sleep now. OK?"

"Well, even if I didn't, if I didn't do just what I said. . . It's just that I know that, something I did . . ."

Memory is mother of all wisdom.
                    —Aeschylus, *Prometheus Bound*

18.

Blind to time, Henry is ever contained inside the bubble of the present. I imagined at first that this blindness would protect him. If he was unaware of the occurrence of any events in his life, at least since some time before the summer of 1953, forty years, how then could he have any sense of what a great empty space his life had been? Whenever you dress him up, he thinks he is twenty-five years old and is being taken to a clinic in Boston for tests. When he hears rock music he thinks it's swing. His mind was sealed off before man flew into space, before the term astronaut was coined; when shown pictures of astronauts he calls them by their 1940s name, rocketeers. Shown a photo of Muhammed Ali in a crouch, he says without hesitation, "That's Joe Louis." In old photos he instantly recognizes faces that the young scientists testing him do not recognize: Alf Landon, Calvin Coolidge, and Hermann Goering. He seems not to know what has happened.

And yet, he knows.

Occasionally over the years he has been queried about his feelings toward himself, and surprising knowledge of his current condition appears there. Questions are put to him in the form of statements about himself: I feel I have failed more than the average person. Yes, he answers. I feel like a complete failure as a person. Yes, he says again. I am disappointed in myself. Yes. I feel inferior to others. Yes. I am feeling useless. Yes. I have a feeling of hopelessness. Yes. Sometimes I feel panicky. Yes.

They might be the replies of a deeply depressed person, except that he is not one. In fact, his answers are accurate. Dark though the image of himself may be, his sense of himself is sharply realistic.

From time to time in the record there appears evidence of his view of himself that is both vivid and troubling. He seems placid, day to day, moment to moment. He seems to accept his situation, and he speaks of his life as a kind of continuing sacrifice: "Well," he says, "whatever is learned is learned, and that's most important." But at times his resignation has failed. When his troubles are confirmed too acutely, as when people tease him or when he is pushed in a way that brings it all before his face, he has reacted fiercely. I have wondered since the time I first heard of his isolation and damnation to the gray place where he now is whether he might hate it, even though he did not fully know it. Does he get angry?

In combing through the records, there it is—amid the prosaic descriptions of doctors coming and going, of their visits to Henry and his mother, of talk of his bowels and his television habits, of the fits when his body twisted and fell (When was the last fit, can you recall, Mrs. M.?)—among these scientific acrostics, on yellowing pages, there is mention of something more alarming, of phone calls in the night.

In the beginning, during the years just after his surgery, Henry stayed at home with his parents in Hartford. There were times when Lillian Herrick, the psychiatric nurse and a cousin thrice removed from Henry's father, with whom Henry would live briefly in the 1980s, came to help out. She said that when she first saw Henry and his mother together, she was struck by the anger in their relations: she would press him constantly to remember for himself what to do and would nag endlessly when he did not do what he was expected to do. She would shout at him; he would take off his glasses and hit her with them.

Henry's mother told stories of his anger as well; she described one day when Henry suddenly became angry with her. He shouted and threatened to run out of the house, as he sometimes had done. He then began to hammer his fist on his bedroom door and continued until the meat of his left hand was swollen, dark red and black, as he cried out, "I can't remember, I can't remember . . ." He broke a finger in the outburst.

Henry's father died in the days before Christmas 1966. Just after that, though Henry appeared unable to remember his death outright, he plunged into an emotional spin. He fell into a series of "automatic motions" that sometimes characterize epileptic behavior before fits. He would rub his hands together faster and faster uncontrollably; he would walk around in a circle rapidly; he would sometimes stir his coffee faster and faster until he stirred the cup right off the table or his mother grabbed his wrist. He kept "acting peculiar," as his mother put it, until finally he crashed to the floor in a fit, with his head jerking back and forth "as if it would snap off."

In another, more alarming, entry, the log reads: "Yesterday, Tuesday, March 24, 1970, while at the Regional Center, Henry went into a totally unexpected and apparently unprecedented rage," wrote Dr. Hans-Lukas Teuber. Actually, unknown to the doctor, there had been other such incidents, but Henry's mother had not described them in detail yet. And there would be others.

On that spring day, Dr. Teuber relates that Henry jumped up in the middle of his work at the Hartford Regional Center, a state rehabilitation clinic populated chiefly by the mentally retarded, at which Henry's work was to test balloons and then mount them on a card. Someone apparently had taken Henry's materials, perhaps wondering if Henry would notice or remember it.

Henry stood, shouting "that he was no good to anyone, that he was going to do away with himself, that he was merely in the way, that he had no memory."

He warned that no one should come near him. When one attendant tried, Henry lashed out and kicked him. "He flung one man clear across the room," the astonished Teuber reports.

The rage went on. Henry said he would kill himself, he would go to hell. Then, he went to the wall and threw himself against it. He bloodied his head as he slammed it with force into the wall repeatedly, "in what looked like a genuine effort at self-destruction," Teuber wrote.

A doctor arrived and injected him with phenobarbital even as Henry bashed his head on the ground, with several men holding him down. When at last Henry slumped into a drug-induced calm, he was taken away to rest.

Dr. Teuber concluded the record with a fearful speculation. Perhaps Henry was beginning to realize what a horror his life was. What would come next? His mother had independently come to this conclusion. She told Dr. Teuber of incidents within weeks of his major outburst that were similar, though not so fearsome.

If things continued to get worse, what would they do with Henry?

At work the day after Henry's outburst of anger, Henry seemed to his colleagues a little nervous. But when he was asked about the day before—Did he remember anything unusual?— he said no, he couldn't recall anything of the day. His mother said the same of his outbursts of anger and self-pity at home: he would very quickly forget them. Despite the intensity of the outbursts and even the injuries he suffered which might remind him, all memory of the incidents appeared to be gone.

Henry does feel anger. Could it somehow accumulate even if he was unaware of it? How would he know? His pale angry figure burst forth at times, mouthing feelings and insights that he seemed nearly unable to express otherwise. But quickly the surface of time ruffled again, and he lost it.

Still, somehow, he fears his own reactions. He asks, most every day, in an idle manner, "What went just before . . . Did I say something wrong?" To me he asked the question in different forms several times. And each time he looked at me expectantly. It seems rhetorical. Does he really want an answer? I hope he does not really want or need an answer, and is at peace, as he seems most times.

Here we find that squares of the mind are stranger than we

imagine; the memory of our own traits does not depend on our collection of personal memories. They inhabit separate provinces. We do not arrive at summaries about ourselves by concluding from instances. Rather, we acquire and store a sketchy knowledge of ourselves more as we store word fragments than as we store the events of our days.

Amnesics, it has been found, do remember shards of what they are like, their own personalities. They have no idea who they are in the world, or the backhand achievement of their own—these greatest patients in psychology who have taught us much of what we know about memory—this is knowledge inaccessible to them. But information about the way they are is there somewhere.

Henry has displayed affection as well as anger, though his fondness, like his fierceness, is short-lived. Few people have ever shown much real attachment to Henry. Apart from the struggles with his mother at home, of course he had much affection for her. There were a couple of times in the record when Henry was away from his mother at M.I.T. that it appears he guessed she might have died. Once, at night, he got on the phone with her and was visibly moved to those nearby. He said with a voice that nearly broke, "How are you doing?—Gee, it's good to hear you—take care—take care . . ." In another instance, after Henry came home from M.I.T., Dr. Teuber noted, "mother and son hugged each other without words for a moment, and Henry stroked his mother's cheeks and shoulders. Both were evidently moved." The only notable affection shown toward Henry was by the janitor at the Hartford Regional Center, where Henry went in the 1970s. In the records the janitor appears at key moments, taking Henry to the hospital after an accident, driving Henry to work and back every day. He is referred to in Henry's records simply as Mr. Bucker, his residence and situation in life unknown. Dr. Teuber called him a "bluff and genial person, a former Army cook," who took Henry away from the crafts table at the center and out into the yard to get him some exercise. He gave him work decent for a man to do, he said. He asked Henry to help with the raking, the painting, tidying up around the machine shop and the boiler room.

He spoke to the curious scientists about Henry from time to

time. He told them that when Henry was inside, they had him putting different-colored balloons into plastic bags, checking that they had no holes and counting them as he went. He could do it, give them a good inspection and keep count as long as he was not distracted, Mr. Bucker said, but he would not remember when to stop—he counted and counted until the bags were stuffed past full; he had to be reminded again and again when to stop. Dr. Teuber noted in the record, with some puzzlement, that people at the Regional Center seemed baffled by Henry's lack of improvement. "We don't seem to be able to rehabilitate him," Teuber describes the staff as saying.

But when Mr. Bucker asked him to come out to the machine shop, things seemed to go better. He asked Henry to fetch tools. "He did it many times, but Henry never came back," the record says. Mr. Bucker tried different tactics. Eventually, Bucker drew pictures of the items he wanted, handed them to Henry, and, given a bit of time, Henry could succeed.

Though Mr. Bucker was not an official person in the record, he may well have been Henry's best friend, from any time in his life. Dr. Scoville had virtually disappeared from the record, after an initial interest. Those from M.I.T. liked Henry, but saw him relatively infrequently, for days or weeks at a time, a few times a year.

Mr. Bucker, the records say, "is 64, obese, and short of breath." And one afternoon, in February 1970, he came by to say that he had been removed from his job as chief of maintenance at the Regional Center. Notes the record: "He says he is a sentimental man, with moist eyes says he loves Henry, wouldn't let him stay idle, wants to make sure Henry feels wanted. If he gets transferred he wants to take Henry along."

He did not, though. He disappeared from Henry's life after a few years and with him went any scrap of knowledge Henry may have gained of him in the years they spent together.

So much of the world must seem puzzling to Henry, as the dimension of time is lost and perspectives are flattened: the four dimensions to which we are accustomed fall into the three which make up space alone. His world has become a room no more than a moment on each side.

Part V

*"He is indebted to his memory for his jests,
and to his imagination for his facts."*
                                        —Richard Brinsley Sheridan

*"It's a poor sort of memory that
only works backwards."*
                                        —Lewis Carroll

19.

I went to work in the days, tended the children in the evenings and weekends. My wife was weakening. We talked of love and of memory, but I found it difficult to make time to pursue Mr. M.'s story. After sampling the history of memory and learning something of Henry himself, I wanted now not broad knowledge but information that would penetrate deeper into the chemistry and biology of memory.

I have read that we emerged from the sea when we were able to take our water with us. We sealed seawater inside the sacks of our bodies to protect us from the poisonous oxygen in the air. I suppose then that we should not be surprised to find that the world inside the brain is more like that of a reef than of a terrestrial garden, inhabited by soft swimming creatures, where wet chemistry is paramount.

The brain was once, in primordial form, an expanse of sensitive skin. It was like the skin of the jellyfish, across which signals from the environment were received as if by antennae, then conveyed to other cells. This suggests something important in philosophy—that the mind is dependent upon sensation for its material to work on.

Unfurled, the brain would be about the size of an opened newspaper and not very firm (indeed, the sheet can melt or leak from its vault). This wet sheet is infused throughout with jelly-like threads, the neurons. In embryo, its evolution is replayed: first neural cells grow in a sheet, then bulge out like a bladder atop the spinal cord, and this bladder grows into the brain's outer layer, the cortex, a word which means "rind" in Latin. This is what we see, the surface of a pinkish gray melon, about three pounds of tiny seas, in a peculiar wrinkly shape. It is not solid within, but is a sheet bunched up, filled with fluids, and set on the tall tube of the spinal cord. The active elements in this odd machine are the infusing threads, "the crown of evolution" as they have been called, the celebrated neurons. They receive signals from outside the body, transmute them in their own peculiar manner, and it is the orchestral summary of their activity which feels like experience.

They operate as shivers of electrical charge run down them, and then, between one thread and another, small chemical tides move. What causes the shiver of charge to run down a neuron's back or to delay that run is—our origins in the ocean speak!—the balance of salts within the neural threads. A small electrical tension between the charged salts inside and outside the neuron is the fulcrum upon which thought and action move: a negative charge within the body of this gangly cell is balanced with faintly positive salts outside it. It remains there, in a slightly uneven balance, like an artist on the high wire waving as he overbalances slightly, until the cell is influenced by the preferences of other neurons. They touch it and convey their own charges to the mix. Some of these neighboring cells signal to open pores and allow the positive sodium salts to seep in, some do the opposite. But the pivotal point is that each neuron is always in a balance, ready to fire if enough signals come from other neurons, or ready to withhold its own firing and thus its own message, if other signals prevail. The tension within the neuron over whether to fire (or in fact, how often to fire) is like that between earth and storm cloud, and each arriving signal from another neuron contributes its preference: release or hold. A great bush of tentacles at one end of a neuron receives preferences and sums them by the hundreds. Depending on the summary a neuron may become hunched in inhibition or stimulated into a rapid quivering. Thus, the shifting salts are sucked in and exhaled, fast or slow; the neuron breathes, and its panting is the sound of thought running.

The neurons, then, are like minute sea creatures, packed side to side like tiny bristles, several hundred billion of them in the whole cranial vault, and each in a frenetic state of decision or indecision. Each bristle has thousands of fine filaments to connect to others, and with the billions of cells, times the thousands of filaments, times the different signals which may pass between each reaching tentacle and another, there are, all told, tens to hundreds of trillions of tender signaling junctions formed among neurons. It is these junctions, the synapses, where neurons pass signals to one another. And at the connections between the cells of thought learning can be shaped. Changing the links—making them highly conducive to messages or making them resistent to messages—can make the neurons of thought either more ready to become active, or more reluctant.

Scientists have created different models to explain the action among neurons that gives rise to different types of memory.

One of the recent models explains how neurons can become permanently altered to create memories. The name scientists use for it is the opaque phrase "long-term potentiation."

Dr. Gary Lynch of the University of California at Irvine is one of those working on the issue. I met him for the first time in a trailer out behind the serious buildings at Irvine. It was after work and he was ready for a beer. His brown curly hair hung low and suggested rebellion. He talked about the history of brain science, and his skepticism about much of the orthodoxy. He spoke, however, with an almost religious reverence of the memory mechanism in which he believes: LTP, long-term potentiation.

Memory, he suggested, consists of an assembly of neurons firing at the same rate or same time as when we first experience an
event. As the same assembly is lit, we "re-experience" the event.
But crucial to retaining this re-experience more or less permanently is the ability of neurons to be ready to join together in assemblies during their firing.

Perhaps a hundred receptors on a single neural cell must receive chemical signals from other neurons. The input from
these hundred receivers leads to the opening of a hundred pores
in the cell wall. The salts rush in when the pores open, and thus,
as the electrical charge inside the cell builds up, the cell fires.

Learning consists of getting neurons to fire more readily on a
second or third try than on the first. Getting cells to fire more
easily depends on priming them: for example, when our rapt attention during an experience suggests the event is important,
and signals to other parts of our brain that we must remember.
The added attention makes a signal stronger, and when a signal
is strong enough—say, 150 signals at the same time instead of
the usual 100—the cell begins to do more than just fire. It begins to open a second set of channels, and the recruiting of
these new channels more or less permanently makes the cell
easier to excite by the same or similar stimulation. After these
changes, it would take only 75 or 50 signals of average strength
from other cells taking information from the world, to open up
the cell pores and set off the neuron.

After some years of pursuing the biochemistry of this phenomenon, early in 1994 Dr. Lynch announced that he and Ur-
sula Staubli of New York University had found a drug which, in
animals at least, artificially mimics the process. The drug holds
the pores of a cell open longer, and so makes it easier for the cell
to receive enough stimulation to fire. The drug mimics moments when we exert enough attention to fix something in
memory, or repeat it again and again to set it in our minds.
With the drug, rats learned mazes in half the time; their neurons were primed for firing, for learning.

This research will ultimately, and perhaps sooner than we expect, lead to the creation of drugs that can enhance the power
of memory. It is unclear how such drugs will be used. Patients

who suffer from Alzheimer's may benefit, but in addition there is likely to be a massive market for such a drug among those who are merely worried about their capacities or, even if they are not worried, seek to gain a competitive advantage over their peers. Among the issues of the future, this question of how cognitive powers may or should be enhanced is likely to be a significant one.

20. ✒

Neurons are at the base of learning, but we must also try to imagine how the data of the senses is handled by the brain. What is the path of data that enters our brain? The neurons, after all, are not randomly packed and joined to one another, but form regions, bulbs, and tiny organs within the brain. Information which appears at the threshhold of the senses is passed along branching paths into the regions interested in what they have to transmit. The threshing of information moves, roughly, from the oldest parts of the brain to the newest, and thus from the simplest types of perception and detection to the more abstract and complex thought.

The oldest area of the brain is the bulb of the brain stem just atop the spinal cord. It is the first stop for fibers running up from the spinal cord and absorbs signals from the periphery where the senses are arrayed like antennae. A flood of impulses to and from this great terminus gives the higher brain parts their "tone and vigor," as Alexander Luria puts it. It is like the inflation of a balloon. "If the influx of these impulses ceases," he writes, "the person lapses into a semi-somnolent state, and then into sleep."

At the next stage, signals are grouped in preliminary bundles:

the chief target of the flood of signals entering from the world is the great hemisphere, the dome that rises above the ears. Signals from the eyes are sent for decoding to the back of the dome, just above the ears; signals of touch from the skin are sent to the top of the dome, just back of the middle of the head; signals from the ears are sent to the side of the dome, just above the ears; signals from the tongue and nose are sent to areas near the temples.

And here, at first, the signals mimic the shapes of things in the world. The cells in some areas of the brain are arranged in actual "maps" along a strip of cortex, where an object's signals form a pattern resembling in some way the object of perception. For vision, a little map of the image that falls inside the eye is recreated by cells on the back of the cortex. For hearing, it appears that the brain maps sounds according to their pitch first. All of these percepts are laid out in some spatial order, for example, with the sensations of one organ lying jowl to jowl beside a related one, and the whole row fanned out in some sensible order: tongue, jaw, gums, teeth, lips, face, nose, eye, thumb, fingers, hand, and so on.

These actual maps are only the first step of processing, in which the form of the world outside is first laid down into the code of electrical and chemical chatter in which neurons speak to one another. We do not have any experience of the signals from the world at this stage, but they are translated into the next iteration of the code, as a code clerk might analyze each map, to transform it into strings of code which no longer resemble the original objects they represent, but are in another logic, another language. The world becomes bursts of Morse code that will be useful to other decoders downstream.

For example, the eyes perceive a broad scale of lights and darks, while the next array of neurons to which this sense data is passed is specialized and responds only to the edges, lines, textures, and curves within the array of signals sent up.

Degrees of brightness, and spots or edges where the contrast is sharpest, have their own arrays of cells awaiting them, like templates or molds waiting to be filled with the promising forms coming in.

Odors carry a signature based on the shapes of the molecules, complex combinations of different basic sets, which move from the nose backward and upward into the area near the temples, where the cascade of feature analysis begins for smell. Pressure-sensing neural cells on the skin react variously to different textures. Other cells sense temperature change (not warmth or coolness itself). Others monitor gravity like a gyroscope, and still others sense body position in space. The neurons controlling heartbeat sense chemicals given off from other parts of the body, where they indicate such things as muscle action or sudden alarm. Other cells respond to hormones like insulin, and some to natural pain-relieving substances (which are effectively mimicked by such masquerading molecules as those of morphine or nicotine).

All of these elements of perception are relayed up to another array of cells, where another round of recognition takes place. This next set of cell assemblies produces a summary from dozens of different, specialized detectors, whose "edges" and "surface feel" and "color" amount to recognition of objects. This set of summaries, in turn, continues to move upward to arrays of cells waiting to detect within or among the signals piped up to them a set of still more abstract meanings.

The projection of the world, to us, is dependent upon these millions of minuscule projectors, which are neural cell assemblies. Because we experience the world this way, we are susceptible to telltale, peculiar errors. These "illusions" are proof that the system works by codes of experience. There is, for example, the persistence of optical illusions, some of which cannot be overridden even by conscious work. Simple working parts can be fooled about what is in the world and what appears to be. Throughout biology such oddities occur. One I find endearing for some reason is the tale of the frog: many scientists have devoted unconscionable amounts of time to the way various creatures, from sea snails to cats, see and understand the world. Among this literature is the article from scientists at M.I.T., led by Jerome Lettvin, which they entitled "What the Frog's Eye Tells the Frog's Brain." The researchers were surprised to find that there are only four major detectors in the frog's eye, meaning of all the mass available to see in the world, the frog was looking for—able to detect—only four general varieties of things. The sustained-contrast detectors notice objects that are distinct from background and remain in position for a time; the moving-edge detectors alert the frog to movements around it, for example the large things that eat frogs and the small things eaten by frogs; the net dimming detectors note the large shadows of ominous objects moving nearby and changing the light; and finally, most interesting, the net convexity detectors become excited only when small dark areas of contrast move across the visual field of the frog.

Flies! As a system for catching bugs and avoiding big hulking things that move, this visual arrangement seems very effective. But of course, frogs have no special apparatus to identify the dark spots after they alight on a leaf. Thus a frog, resting on a heap of freshly killed flies, may well starve.

The slim telegraphy we can capture from the air nevertheless breaks across bestial borders like waves, presses through our membranes like ghosts through walls. The cells are raised in choruses of sight and sound and scent; air bats the skin drums of the ear; light pierces the wet glass of the eye, setting alight the fuses arrayed there. We have a number of detectors, where the frog has few, but set beside all of the possible things to see, we are blind.

Memory itself is the step beyond the interpretation of fresh data from the world. Once such data has meaning, that is, a place among our previous arrays of experience and deduction, it is sent for further bundling. In the center of the head between the ears, the mediators of the hippocampus and its satellite organs absorb the recognized objects, the symbolic meanings which have been suggested by another bank of cells, and the emotional tones added en route by limbic areas of the brain; then all is consolidated, bound for permanent storage elsewhere.

(The bulbous parts of the middle of the brain are striking partly because they are ornamentally elaborate. The hippocampus, curled like a sea horse—or a pair actually, as there are duplicates on the left and right, as with most of the rest of the brain's parts—and the amygdala, the "almond," lying beside the inner cheeks of the sea horses. The striped layers, "corpus striatum" in anatomy, lie near the bulbs of olfaction.)

It is at the hippocampus that objects and meanings already formed in previous stages are sorted and tagged, and those things to be remembered for more than a few minutes are sent to storage sites. (The places of long-term storage are, paradoxically, back up on the dome at sites near those where the fresh data first came in.) Whole scenes only emerge into experience, with their emotional and intellectual meanings, when the waves of analysis of the different hierarchies within the dome bring the disparate parts together in a river of experience moving through the hippocampal area, and through the front of the cortex where plans and symbols are sorted.

The final places of storage and the place where fresh data first arrive are spread out across the cranial dome, and are what scientists refer to with a wave of the hand and the phrase "association cortex." Here lies knowledge; once received and processed, it is joined here by some unknown scheme to all the rest of what we know and feel. Thus, when we draw up memory and knowledge, it comes up as a net from the water, with not only the piece we intend to draw up but the many knots in close association to it. The cerebral cortex, as Dr. Richard Thompson writes, is what makes humans human, and within it lies a critical part of the secret of human consciousness. It is this part of the brain which has increased most in size as human crania expanded over the past million years, and it now occupies a far larger percentage of the brain in humans than in other animals.

In thinking about thought, it is difficult to imagine all this signaling as experience, difficult to relate it to what the world feels like to us. One question, however, which helps clarify these matters is, what would occur if there were crossed wiring in the maze of sense circuitry? If the cells that indicate that an apple is seen were somehow, in a Frankensteinian operation, cut and rewired to the cells which do not experience "seeing" but instead the ones that experience "feeling," what would happen?

There is such a phenomenon, synesthesia, in which the signals from one sense are involuntarily experienced by more than one assembly of cells at the same time, like the patient reported by Dr. Richard Cytowic. The young man explained to him that flavors have shape. Thus, when explaining that the sauce he had made for a chicken was disappointing, he said, "I wanted the taste of this chicken to be a pointed shape, but it came out all round. . . . I can't serve this if it doesn't have points." He actually feels tastes, mostly on his hands and face, but the sensations can sweep over his body. A chocolate pie is not only chocolate, but a series of cool, smooth columns.

There are a great many such accidentally mixed experiences named in the medical literature. Some of the strangest include the sudden multiplication of images after seeing them, or the sudden change in size of objects to very great or very small, much as they occurred to Alice in her adventures in Wonderland, as Cytowic writes. And among these is the "out of body" images in which we apparently look down upon ourselves, an impression favored by those telling of after-death experience.

Vladimir Nabokov wrote of himself that he presented "A fine case of colored hearing. The long *a* of the English alphabet. . . . has for me the tin of weathered wood, but a French *a* evokes polished ebony. This black group also includes hard *g* (vulcanized rubber) and *r* (a sooty rag being ripped). Oatmeal *n*, noodle-limp *l*, and the ivory-backed hand mirror of *o* take care of the whites . . . Passing on to the blue group, there is steely *x*, thundercloud *z*, and huckleberry *k* . . . in the green group, there are alder-leaf *f*, the unripe apple of *p*, the pistachio *t* . . ."

Much of the elaboration of brain and memory and even consciousness also exists in animals, if we exclude our own self-consciousness or reflexive thinking. Other animals have developed mentalities that may be much like those we see in dogs now: capable of feeling, capable of solving elementary problems, having some sense of what's more important and less important, moment to moment. In the world of the dog, social status, food, and sex are essential issues on which they work moment to moment. But there is no explicit memory of things, no mental representations of the abstractions "status, food, sex." The memories of animals are quite complex, though not full of images or past pictures. They have stored memories of thousands of distinct smells, each of which is associated with needs and behaviors, desires and aversions.

As Elizabeth Marshall Thomas describes a dog weighing alternatives: "something off the trail had drawn the young female away from the group, so that when the others stopped at the river, she wasn't with them. They turned to go home, as was their custom, and had gone about eighty feet down the trail when the young female, hot and ready for her daily bath, burst out of the bushes halfway between them and the river. Too late— they were leaving, and she had missed her swim. Poised beside the trail, she first looked to the right after the group, then looked to the left at the river, then looked to the right a second time, then looked once more at the water, made an instant decision, rushed full speed up the trail to the river, plunged in, quickly swam a few strokes, and then turned back to the bank, leaped out, and tore after her group, not stopping to shake until she had caught up to them."

This is intelligent behavior of some sort, and requires memory, though the thoughts in it lack concrete expression. There are no words or images, and thus it cannot be communicated very explicitly, either. The dog has "categories" which it forms, such as the species of different animals, whether they are food or predator, what a female or male is, what a pup is and whose it is. These are categories of recognition. But dogs do not have a mental grammar by which these categories can become manipulable concepts like words. They can string together associated thoughts, but cannot use these thoughts independently from the things in the real world to which they refer. They cannot become symbols. They cannot realize, for example, that a dog is in some way like a young human, that a cup is like a pair of hands, or that two-dimensional sketches can represent three-dimensional worlds.

Oliver Sacks once said in an article on vision and blindness that even the seemingly automatic skill of seeing must be learned, and must be exercised. "When we open our eyes each morning, it is upon a world we have spent a lifetime learning to see. We are not given the world: we make our world through incessant experience, categorization, memory, reconnection." He told of a patient, Virgil, whose sight was restored after forty-five years of not seeing—but at first, he could not understand the world: "He experienced lights and darks, color and line, shape and depth, but all in confusion. They were not coherent objects in a coherent space. His first glimpse of the world was of his surgeon. No cry (I can see!) burst from Virgil's lips. He seemed to be staring blankly, bewildered, without focusing, at the surgeon, who stood before him, still holding the bandages. Only when the surgeon spoke—saying 'Well?'—did a look of recognition cross Virgil's face. Virgil told me later that in this first moment he had no idea what he was seeing. There was light, there was movement, there was color, all mixed up, all meaningless, a blur. Then out of the blur came a voice that said, 'Well?' Then, and only then, he said, did he finally realize that this chaos of light and shadow was a face—and, indeed, the face of his surgeon."

He remained confused about some things. He saw birds—they made him jump sometimes if they came too close—though his wife said that of course they did not come close at all, but Virgil had no idea of distance. Sometimes surfaces or objects would seem to loom, to be on top of him, when they were still quite a distance away; sometimes he got confused by his own shadow—the whole concept of shadows, of objects blocking light, was puzzling to him—and he would come to a stop, or trip, or try to step over it. Steps, in particular, posed a special hazard, because all he could see was a confusion, a flat surface of parallel and crisscrossing lines; he could not see them (although he knew them) as solid objects going up or coming down in three-dimensional space. The idea of painting, representing three-dimensions in two, was baffling."

Earlier, Richard Gregory had a similar experience with the patient S.B., born blind but who gained sight at age fifty-two. At first he could recognize little or nothing, but within some days he was able to "see" objects which he knew by touch; those of which he had no experience startled and puzzled him. When blind, he could work a lathe; when shown it in person he had no idea what it was and had trouble with "seeing" it; it appeared as a confusion of lines and angles and odd bits. But after he closed his eyes and examined it with his hands, he could see it. I suspect, though, that Mr. B. never did see the lathe as we might. I would guess that the parts he knew well, and his understanding of the machine, established so long, dominated his "image" of the lathe, just as in the famous experiments when rich children are found to perceive the size of money as much smaller than poor children perceive it, and A students perceive their teacher as shorter than the failing scholars, who draw their teachers as quite tall.

21. ✍

Now, we can place the disability of Mr. M.: he can see, perceive, decode; he can recognize objects and their meaning; he can manipulate them in thought, and react to them with sensible behavior, laughter, or irritation. But when all converges on the hippocampus, it runs off the rails. The signals bleed off the ends of neurons, shorn like telephone wires on their way to the receiver. He cannot hold anything for further thought or send it back to be permanently stored at the sites on the dome. What was formed and laid up there before his surgery, of course, may still be brought down into conscious use, into the front of the dome where planning and conscious manipulations of thought occur, but nothing new may be caught from the stream of experience.

The accumulated experiences lead steadily in one direction: they create what may be described as a full-feeling representation of the world within the vault of the cranium. Like models sculpted in clay or topographic maps on paper, these representations are unlike the terrain they depict in most respects. The slope of a grassy hill is, on a paper map, represented merely by pale green background with black lines spaced proportionally—

closer together suggests a steeper grade and farther apart marks a more gentle drop. So it must be in our internal representations of the world: we navigate not through the world, but through our own approximations, outlines, caricatures of it represented by the familiar fireworks of the cortex. Thus what we see and feel as the world is not really the world, but our inner model of it, as if we were pilots in a virtual reality bearing on a general likeness to the landscape through which we fly. Henry has formed these maps, during his twenty or so years of unhindered experience as a child and young man in Hartford. But a person with no hippocampal system from birth, if there were ever to be one, would have not even this.

To make some connections between biology and experience, I had flown out to San Diego to talk to Dr. Larry Squire. Squire had begun to study memory in the years just after the story of Mr. M. appeared in the scientific literature. A contemporary of Suzanne Corkin, he became absorbed in similar issues but from a somewhat different, more anatomical perspective. Cells and fields of cells, linked like rope bridges across the brain, were what fascinated him; how were these the underlying structure of memory?

I had spoken to Squire on the phone at some length and found him able to navigate the streams of English and acad-emese fairly readily; he also had the ease and openness of a Mid-westerner. I thought to myself, from him I might learn something. His office was tiny, as are most active scientists' quarters, and he stood a little awkwardly, dodging stacks of pa-per on a desk as he shook hands. He wore blue corduroys and a striped shirt, the requisite rumpled look. He had a short, stubbly beard, and little, round metal-framed glasses. On the wall hung a plaque—the regional Scrabble championship. He says that Scrabble is not really a memory task, and anyway, he doesn't have time for it anymore.

I sat knee to knee in his office at the Veteran's Administra-tion Hospital as he explained, with a complete set of gestures and verbal signs of excitement, how hippocampal cells are linked back to areas thought to be sources of emotion and for-ward into areas thought to perform analysis and planning. The hippocampus is a binding area, where disparate threads of experience are linked and somehow bundled into more or less permanent memory units.

The meaning of Mr. M. and the work of Brenda Milner, then Suzanne Corkin, did not really dawn on researchers until well into the 1970s. The work was begun immediately—even Dr. Scoville became involved in efforts to mimic, in monkeys, the amnesic syndrome seen in Mr. M. But at first, it didn't work. It was some time before Dr. Mortimer Mishkin at the National Institutes of Health combined the skills necessary to sort out which behavior in monkeys was similar to that which showed memory in humans. By 1978, Mishkin had demonstrated memory impairment in a monkey with a cut in the brain that essentially mimics that in human amnesics. The cut was in the hippocampus.

But in the beginning, "it was real confusing," said Squire. "Now this is a reconstruction, but my own sense of it is that we really didn't understand until very recently what the human amnesic syndrome was. That is what we were trying to model."

"Brenda Milner had showed, in about 1962, that H.M. was pretty good at learning what she called motor skills. By motor skills she had in mind tasks like the ability to trace in a mirror a drawing of a star. It's like first combing your hair in a mirror: at first you're very clumsy, and then you get better, and then you take it for granted.

"Well, H.M. showed a day-to-day improvement in that. Then what happened during the 1970s was the realization that there were other kinds of learning and memory, aside from motor skills, that amnesic patients were normal at.

"Patients who could not remember, they discovered, were able to remember some particular things. Memory was not a unit, but was a number of separate abilities."

There are a few peculiar tasks that psychologists make patients do which seem to give a clearer idea of at least two distinctly different sorts of memory.

One of the little exercises is called "priming." A patient may be given a list of words to read. "Do you know these words? Would you rate them on a scale, one to five, which you like the best?" the experimenter asks, trying to be the sly scientist, asking questions he doesn't want the answers for. He just wants the subject to pay a little attention to each word. "Motel. Absent. Income. . . ."

After a few minutes of other work to distract the subject, the subject is given incomplete words: "MOT . . . Now just say the first thing that comes to mind. Allright, now ABS . . ."

Many words can complete these stems. Motor. Mottle. Motive. The psychologist must do his work properly, disguising the tasks by, say, mixing the stems that are the target in with many stems the subject has never seen before, making sure the subject does not think he is trying to remember the words of the previous list in some kind of memory exercise. The subjects do not consciously recall the words from the list they saw some time ago, but nevertheless, they very frequently respond with a word from the list. They blurt out "motel," not "motor."

The result seems logical enough. They are "primed" by the previous appearance of the word, primed to say it again more often than digging down for another word with a similar beginning.

"I really don't know what's going on precisely," said Dr. Squire, "but the way one would talk about it is to say that the presentation of the word 'motel' activates the preexisting representation of that word which we have in our brain somewhere, presumably in the neocortex of the left hemisphere. The word is then temporarily activated, and during the time it's activated it's there to influence your behavior—you see a part of the stimulus, M-O-T, and you'll blurt out 'motel.' Or if someone asks you for an association, such as 'What's the first thing that comes in mind when you hear the word hotel?' You might well say 'motel.'"

I asked, "Like being in a conversation, in which the context will determine the meaning of words that might be ambiguous otherwise, but that are not ambiguous in the conversation?"

"Right. Yes."

What all this suggests is a level of unconscious processing of words, among other things. It is "memory," but it is not recall like Proust's, in which he recreates a moment and its setting.

Rather, this is a paler, mechanical kind of memory. Another mechanical, or as Squire calls it, procedural, kind of memory is "source" memory. I may recall that a quotation is on the lower half of a right-hand page, but not remember either the page or the exact words of the quotation.

Many abilities fall into this "unconscious memory" category. Some are physical skills like playing the piano, others seem more intellectual, like recognizing patterns in a chess position without consciously going over them in detail.

The easiest examples Squire cites are motor skills. "If you learn a tennis stroke, say a backhand, and I ask you, 'What is it that you've learned?' You say, 'Well, I've got this backhand.' The knowledge is embedded in the procedure you have for executing the backhand. Unless you're a tennis instructor, you don't have explicit knowledge available about what the skills are in a backhand. You have quite simply to perform it. That's the idea of 'procedural' memory."

Contrast that, he said, to the tennis player recalling encounters she had as she learned the backhand, or the courts where she used it in winning particular points. That is conscious memory, of episodes.

I told Squire I was having some trouble thinking of the memory for a word like motel as a "procedural" or "skill" memory. It seemed, intuitively, a different sort of knowledge than tennis playing. And it is, in that one is a motor skill and the other is a perceptual skill, but both are unconscious learning and neither can be dissected by the learner.

"You were suggesting that priming was one of those things that's procedural. I wouldn't have guessed that at all, " I said.

"No. We wouldn't have either. It was a complete shock to us. If someone had said to me ten years ago, 'Here's this task called priming I'd like you to try out. How would an amnesiac patient do on that?' I would have said, 'You've just got another way of measuring memory there.' But what seems to separate it from other memory is even though you've activated your representation of the word 'motel,' the fact that you're about to blurt out the word is something you are not aware of at all."

Memory thus is scattered among discrete abilities, some conscious, some unconscious. The hippocampus is the organ of memory for some, but not all of them. So it was discovered how Mr. M., profoundly amnesic as he is, is still capable of some kinds of memory.

When Mr. M. is tested on motor-skill memory, or priming, or some other perceptual and unconscious skills, he performs quite normally. And also with short-term memory—the reason why Mr. M. can hold the present situation and facts in mind, briefly. Memory was shown to be several separate engines, each operating together but independently, and each one capable of being destroyed or altered separately from the others.

The hippocampus begins to create memory; it binds together all the elements of a sensation or of a moment, not instantly, but after you've been distracted from it. When your attention moves away, the hippocampus begins to work.

"I'm not sure I understand exactly what the hippocampus is supposed to be doing with all this information. You talked about it in terms of packages of elements, so I guess you imagine that a visual stimulus has different elements, and the hippocampus has to bring them together to make a single image in memory?"

"Well, a first step is that a memory has a space-time context to it. We not only remember this object, but we remember that we saw it in that room. In other words, you not only have to identify what it is, but you also have to identify where it is in space."

"The hippocampus is involved in connecting all those things?"

"I guess," Squire said. "It's unclear what exactly it's doing. We talk in terms of the hippocampus providing some way of linking all the elements of a total representation, somehow fixing it over time. Binding it," he said.

The essence of this kind of memory—wrapping together all the parts of a moment—is that it can fix information all at once. Each episode in your recall of the events of your life is of something that happened once, and once only. In this fact, both the poignance and the biological import of human memory begin to come clear. Animals may learn things after repeating the experience many times, or may learn to react to a smell after only one time. But in human memory the full picture and all its attendant meanings can be reproduced, at once, at will.

"This is what the hippocampus allows you to remember," Squire says. "Without a hippocampus, I think, you can learn associations and fragments; it can affect your behavior like priming does, change your attitude about an animal or something. That could all go on without the hippocampus, cumulative changes in memory storage may take place. But not these single bursts in which all these things come in at once."

22. ⤳

Bees can recall enough of a visit to a flower to signal its location by dance, a specialized power which does not become the power to recognize and convey other information about the terrain. Frogs' eyes can spot flies, as their ganglia can recognize little spots of moving light-dark contrast and link the sensation to "food." But it is in accumulating and linking explicit memories that humanity's mental leap must have taken place millions of years ago. The little memories of animals, bound into groups and categories of groups, allowed humans to break the bonds of the literal world. They could create and transmit elaborate culture. The future was invented.

The more profound the memory loss in amnesic patients, the clearer it is that memory is the central trait of the human mind and that these people have been sent in some way back through evolution to a simpler, less comprehending state. Filmmaker Luis Bunuel lost his mother to Alzheimer's disease, and wrote, "You have to begin to lose your memory, if only in bits and pieces, to realize that memory is what makes our lives . . . Our memory is our coherence, our reason, our feeling, even our action. Without it, we are nothing. . . . we can only wait for the final amnesia, the one that can erase an entire life, as it did my mother's."

Daniel Schacter of the University of Arizona recalls taking an amnesic patient out of the psychology lab onto the golf course; he was curious to see the amnesia operating in the natural world. The man, fifty-eight years old, was a fair golfer before his diagnosis of Alzheimer's disease. When he got on the course with Schacter the man was able to follow the rules of golf. He picked his clubs correctly, teed up in the right spot and in the right way, and hit well. He had not forgotten the jargon of the game, casually and correctly using the terms bogey, divot, duffer, handicap, and so on. But through all this, the memory deficit was startling. He attempted to tee off repeatedly, forgetting that he had just shot. On one hole he hit a good drive right onto the green of a par-three hole, and was pleased about it. But he immediately walked out to look for his ball in the creek on the left of the fairway, and was astonished when Schacter told him his ball was on the green. By evening, the man also was unable to recall any of the day's events, any good or bad shots he had made, and he soon denied ever having played the game that day.

There even appears to be something like a memory for social behavior. Jordan Grafman of the National Institutes of Health told of a patient who had lost the memory of basic etiquette and social rules. The man behaved like a child in social situations. For example, Grafman said, the man once went to a funeral and without embarrassment talked animatedly and laughed out loud, as if he were anywhere but at a funeral.

After I spoke to Dr. Squire in his office in San Diego, he unwound his long frame from the chair, and led me down the hall to the elevator to show me the lab.

In the basement he showed me an array of photographs of brain tissue magnified that belonged to a man called R.B. He had a stroke in 1978 at age fifty-two, and the cells which died out in his brain, starved of oxygen it appeared, were confined to his hippocampus. The result was a memory loss which demonstrated what we had been speaking of. Along a high shelf in the lab were jars; there rests the brain of R.B., amnesic successor to Mr. M. Looking at the jars I thought for an instant of Luigi Galvani, Camillo Golgi, and the obsessive and beautiful drawings of Ramón y Cajal.

The information from Larry Squire in California and Mort Mishkin at the N.I.H. in Maryland helped sort out the different kinds of memory that have been found to be distinct since the case of Mr. M. established that there were different types. There will be many subtypes as these systems are elaborated in detail in the future (some researchers now count ten different memory engines, which do somewhat different things and operate more or less independently within the greater engine of memory and mind), but there are at the moment believed to be three major categories of memory, handled in three different ways by the brain.

One kind of memory which is shared by all animals is the ability to rehearse and learn sequences of skills. These may be perceptual, such as tracking the approach of an oncoming car or following a ball in flight. They may be physical routines, such as typing, shifting gears in a car, or, in a dog, the ritual of going to the door, pawing it, and looking to the nearest human for aid in going out. Some are complex sequences, for example, the ground stroke in tennis, which is a sequence of multiple tracking routines and physical responses. Laid upon one another, routines like these can account for a substantial part of the daily repertoire of all animals.

On the matter of skill memory, musicians speak of their fingers developing lives of their own, as does David Sudnow in "The Ways of the Hand." He writes first of mastering the procedures of playing, and only after mastering, guiding the bundles of learned patterns as they are produced:

"I see my hands as jazz-piano player's hands, and at times, when I expressly think about it, one sense I have from my vantage point looking down is that the fingers are making music all by themselves . . . My hands have come to develop an intimate knowledge of the piano keyboard."

The learning of skills, as researchers found, is handled differently from fact learning. H.M.'s first teacher, Dr. Milner, found this when she gave him a routine neurological test; reading and writing words while looking only in a mirror and not at the paper from which you are reading or writing—the hand must trace words upside down and backward. It requires both the memory in the centers controlling the hand and in perceptual areas for the viewing and the tactics employed. These Henry did learn, rather quickly, and soon became proficient enough. But, having learned the skill, he could not and still cannot remember that he learned it or know that he knows it. It is completely inaccessible to him unless he is reminded and tested.

There is also the Tower of Hanoi puzzle, a difficult little trick that Henry surprised researchers by learning. It comprises three posts. On one is a set of five rings, each smaller than the one below it, so the effect is of a pyramid of doughnuts. The game is to move the rings, one by one, from the far left post to the far right post. But—and herein lies the trouble—a larger may never be put on a smaller ring. Most people take scores or hundreds of moves to succeed at first. But gradually, the number comes down, closer to the ideal thirty-one placements. Henry began the same way, and indeed, he was slow. It took hours, days, of game upon game. But he did learn it and produced several games of thirty-one moves. If he made a few errors, he might never finish a game, but when he started well, he could move through effectively. This too is memory, although not episodes from his life or facts.

All this is skill memory. In higher mammals, apes and humans especially, another sort of memory has developed beyond the layered, routine skills. It is the kind we are more used to calling by the name memory, and it was termed "episodic memory" by Endel Tulving when he classified memory types in the 1970s. It is the recall of specific episodes, whole events, from life. It includes their time and place, along with some richness of texture, smells, colors, images, or emotions.

Henry, though he has learned mirror-writing and the Tower of Hanoi, has no memory of learning them, no memory of the moments when he sat to learn them, and he would deny that he has the new abilities. What he is missing is episodic memory of the type that Vladimir Nabokov in "Speak, Memory," sketches: "I see the awakening of consciousness as a series of spaced flashes, with the intervals between them gradually diminishing until bright blocks of perception are formed, affording memory a slippery hold." He recalled his earliest sense of himself as a person during an episode at his parents' country estate: "Judging by the strong sunlight that, when I think of that revelation, immediately invades my memory with lobed sun flecks through overlapping patterns of greenery, the occasion may have been my mother's birthday, in late summer, in the country, and I had asked questions and had assessed the answers I received. All this is as it should be according to the theory of recapitulation; the beginning of reflexive consciousness in the brain of our remotest ancestor must surely have coincided with the dawning of the sense of time. Thus, when the newly disclosed, fresh and trim formula of my own age, four, was confronted with the parental formulas, thirty-three and twenty-seven, something happened to me. I was given a tremendously invigorating shock. As if subjected to a second baptism, on more divine lines than the Greek Catholic ducking undergone fifty months earlier . . . I felt myself plunged abruptly into a radiant and mobile medium that was none other that the pure element of time. One shared it—just as excited bathers share shining seawater—with creatures that were not oneself." Perhaps the most famous of all moments of episodic memory is Proust's epiphany which accompanied tea and toast.

This sort of memory includes an element of conscious awareness of time and space, the power to draw together all the elements of a scene into a single representation of some kind. It is believed that episodic memory exists probably only in birds and mammals. Apes have rather advanced abilities of episodic memory, because of the complex social world in which they live and which other animals cannot register.

The reason apes can be trained to learn hand signs and other symbols, writes Canadian psychologist Merlin Donald, is "they are using episodic memory to remember how to use the sign; the best they can manage is a virtual 'flashback' of previous performances." They do not use language inventively, creating expressions needed for any situation, in a fluid expression. Rather, they always sign as if they are referring to some unseen picture of things and actions, "Koko-banana-eat" or "Roger-tickle-Washoe." It is this power that Henry lost with his surgery, as it appears that the hippocampus is essential to tying together and storing episodes from life. Though he may still bring back, in a ritualized and truncated manner, some few memories from the time before his surgery, since 1953 no new events have happened in Henry's world. And even when he exercises the more primitive ability to learn skills, he is unaware that he has learned them.

Beyond the skill memory and the memory for episodes, there is a third type of memory which Tulving described. He called it semantic memory—the memory for facts such as names, numbers, ideas, and other similar material we manipulate in daily life. Feelings are often said to be the center of our humanity, but a closer look will teach us that it is rational facts, the symbols and their manipulation, which distinguish us among animals. After all, every animal has fully developed feelings; it is only rational thought, factual knowledge, which is denied them.

Facts are abstract units with no hold in the world, no meaning for other species or in the passage of geological or astronomical time. They are signs representing a secret knowledge of the world which we have made, and the ways we may move it for our own purposes. We are swimming in symbols. The imaginary boundaries of thought run through all we do. In the book of grammar, the abstractions multiply: "The future perfect tense is formed using a past participle ... and describes an action or state of being to be completed in the future before some other action or state of being." The book of history states "Leonardo was born outside the small town of Vinci, near Florence, in 1452. The traditional date of his birthday is April 15." The birthday is hardly worth remembering, but the year is, for its remoteness remains a constant surprise. The entomologist writes "We had observed that the sensitivity of a fly to sugar increased nearly ten million–fold from the time he was fed until one hundred hours after feeding. In other words, as he became hungrier he would respond to more and more dilute sugar so his sensitivity was a good measure of his state of hunger. One hundred hours, however, is a long time to wait, so we devised a method of making flies hungrier quicker." Factual memory is another type which Henry has lost the ability to form.

Together, skill memory, episode memory, and fact (or semantic) memory are only very broad classes and include many quite separate abilities under each label. For example, under skill memory there seems to be a separate system for remembering the time or place of an event, separate from the memory of the event itself, and there are also separate systems for the "look" of words on paper, the shapes of letters, beginnings of words, apart from the meanings of the words themselves. These separate under-abilities in the mind certainly number in the dozens, perhaps more than a hundred distinct or somewhat separable powers. The array of separate engines and stores of memory has grown since their separateness was first discovered in Henry. But one can think of all of them in the schematic: "Knowing How" for skill memory, "Knowing Of" for event memory, and "Knowing That" for semantic memory. They are joined together only in experience. Proust recalled the tea and toast and felt that the experiences of color, shape, and a thousand separate associations were not held together in his mind, but separately, and joined only in the act of recall. The apple is not an apple in our minds, but is its facets—color here, shape there, texture in another place, while smell and taste and other associations may be found as widely scattered as the colors in an impressionist palette. They are joined only by the viewer during experience, as Proust said of the colors of Monet's *Water Lilies:* "We are there . . . trying to chase away all thought, to understand the meaning of each color, each one calling up in our memory past impressions; these impressions are associated in an architecture as airy and multicolored as the colors on the painting, and they build a landscape in our imagination."

We may sort humans from other animals by their use of memory. Each animal has its memory fixed on different central points. Rats have the capacity to recall thousands of smells, with recognition after only one presentation. Humans may know only a fraction of that. Instead, they have become fixed on symbols—each human knows tens of thousands of words, and may manipulate them in a limitless variety of ways to make sense.

"Since humans and nonhuman mammals, including apes, differ so fundamentally in the types of memories they can retain, it is possible to use this fact to characterize their two types of society," writes Merlin Donald. "Most animals, including humans, possess procedural memories, and therefore the term is not particularly useful in characterizing the dominant cognitive feature of mammalian culture. Episodic memory is probably unique to birds and mammals, forming the basis for Oakley's definition of rudimentary consciousness. Humans possess both procedural and episodic memory systems, but these have been superseded by semantic memory, which is by far the dominant form of memory in human culture, at least in terms of the hierarchy of control. In contrast, episodic memory is dominant in most mammals, including apes. Animals do not seem to possess the systems of representation that would allow them to have elaborate semantic networks. Their experience, in this light, is entirely episodic. The pinnacle of episodic culture, the culture of the great apes, marked the starting point of the human journey."

These types of memory also follow up the spiral of evolution. Information is retrieved by the senses, and passed through the brain stem, medulla, and thalamus. Without these there is no wakefulness, no arousal, no consciousness. In some way, this is where attentive creatures leave plants and bacteria behind.

Above the brain stem in evolutionary development is the brain core, the "limbic" (margin or edge) brain, as it was first called by Paul Broca, because it lay at the edge, atop the brain stem. It's better called the "mammalian brain" because it is what mammals have in common; it comprises the organs which are responsible for the regulation, on the one hand, of physical processes such as eating, sexual arousal, and smelling, and on the other hand, the organs responsible for selecting what to send on to the higher brain and for linking together aspects of experience in memory, as the hippocampus and associated structures lie here, finding "matches" between experience and memory, and bundling disparate parts of experience into memorable wholes.

In the dome above these is the cortex, which appeared in animals very late in evolution, at 50 million years or less, which is after 95 percent of evolutionary time had already passed from the human perspective. It is here that the "left" and "right" brain usually specialize into, respectively, logic, language, and

mathematics on one side, and spatial, visual, and artistic thinking on the other.

We can now understand what the doctors were seeing, for
the first time in the history of medicine, in H.M. in 1953, a case
of inability to form new memory: a pure and complete instance.
In each other case, there is some complication that makes matters difficult—Korsakov's syndrome in which alcoholism has
damage multiple parts of the brain, or syndromes in which the
frontal lobes are damaged as well as the hippocampal memory
systems.

With damage to the hippocampus and the nearby brain areas
alone, as Dr. Oliver Sacks has written of another patient, the
ability to acquire information about new facts and events is dev-
astated. There ceases to be any explicit or conscious remembrance of these.

Curiously, it is still possible for a person with such damage to
have unconscious or implicit memory. That is, memory which is
expressed in performance or behavior, but is unknown to Mr. M.
He may learn a skill, use it expertly, and at the same time deny
he has ever done it or been taught it. His good performance is a
puzzle to him.

Such implicit memory allows him to become familiar with
the physical layout, from walking it over and over, and make
judgments about whether some situations or persons were
pleasant or unpleasant. This kind of memory is a sort of primitive under-layer of memory.

The most famous anecdote about such unconscious learning
is that of Dr. Edouard Claparede and his amensic patient. In order to illustrate the effect to a medical school class in 1911, Claparede "somewhat cruelly," as Sacks observes, had a pin
concealed in his hand when he shook hands with the amnesic
patient. The stab of the pin was, only a few moments later forgotten by the patient, but he nevertheless refused to shake
hands a second time with Claparede.

Explicit, conscious memory, however, is more advanced and
requires the construction of complex percepts. It draws on syntheses of representations from every part of the cortex. They are
brought together in a contextual unity, or "scene." Such full

scenes can be held in mind only for a minute or two, the limit of short-term memory, and after this they will be lost unless they can be shunted into long-term memory.

This higher-order memorization, fixing of scenes and other complicated sets of facts or perceptions, is a multi-stage process, involving the transfer of perceptions or syntheses of perceptions, from short term to long term.

All of us on occasion lose something from short-term memory, forgotten what we wanted to say—"but only in amnesics is the precariousness realized to the full," Dr. Sacks says. They can no longer transmit experiences from the present, the short-term memory, into permanent ones.

Where does this leave Henry, with the missing parts of his core mammalian brain but his intact higher and lower regions? Perhaps somewhere along the grade between dog and ape, I suppose, with the power to speak and think, but without the power to record new episodes.

With this in mind, I began to wonder about the elements of memory and consciousness: How do we differ mentally from other creatures, and what do we share?

All the dignity of man consists in thought . . .
But what is this thought? How foolish it is!
The mind of this sovereign judge of the world
is not so independent that it is not liable to be
disturbed by the first din about it.
The noise of a cannon is not necessary to
hinder its thoughts; it needs only the creaking
of a weather cock or a pulley.
Do not wonder if at present it does not reason well; a
fly is buzzing in its ears; that is enough to render it in-
capable of
good judgment . . . Here is a comical god!
O most ridiculous hero!

—Blaise Pascal, *Pensées*, 1660

23. ❧

We were out in the sun once, Henry in his wheelchair and I beside him, waiting for the taxi to take us for his brain scan. It was the usual changeable, disturbing weather of a Boston spring, but just now, it was bright and warm. "Great day!" I said. "And sunny!" As I said it, a shadow crossed the walk, and the sun dived into a cloud. Henry laughed. "Well, just as soon as you say it, it isn't!"

Across the street was a construction site. We watched at length; the crane—it must have been ten stories tall—swung out over the deep hole and back up to its bank, a huge bucket of gray muck gliding down. "I bet they are glad they don't have to haul that all the way up," said Henry. I glanced up the street for the taxi, and quickly Henry's gaze followed. He wasn't sure why we were looking there, and he studied me. My head was turning back to the construction, and Henry's gaze settled in there again, too. He watches and he listens for clues, for the implications of a question, for hints at what the subject is, how he should feel, and how he should answer. How else could he be than like a dog, waiting expectantly at the door? I imagine him walking, always a little uncertain, but compelled to press ahead while around him is a blank fog. "And I moved forward," said poet W. S. Merwin, "because you must live forward, which is away from whatever it was that you had, though you think when you have it that it will stay with you forever."

I recall one of my first visits to Henry. As he talked with Dr. Corkin, who had come to get him for tests, I observed in silence. When she approached, he looked up, blank at first. I could almost put words to the passing expressions on his face: Ah, a face that seems familiar. To talk to me? Yes—she takes up my eyes.

"How are you, Henry?"

He groped a little, feeling just behind him for something. "Fine, I guess," he said, and smiled a little. Again he watched, expectantly.

"Do you know what we're going to do today?"

I felt him turn metaphorically to search for an answer. Then he shrugged. "I don't remember."

But then she handed him his walker. He can grasp it, flip out the legs, set it just ahead of himself, and lean up into it.

"Why do you use a walker, Henry, do you know?"

A brief look into the fog. Nothing there. He looked down. "Well, it's my legs," he said. He quickly realized the humor in this too-obvious reply, and grinned.

Down the hall, he turned left, heading for the experiment room. How did his body know to go that way? Part of his brain has learned, though the other has not. The one that is supposed to keep track of what has been learned is missing.

At the plain gray table in the experiment room, Dr. Corkin asked, "Do you know where you are, Henry?"

Again, he looked out into a fog. But here! There is something! "Well, at M.I.T.!" He beamed; he gathered a scrap for his questioners; he always likes to please. It has taken decades of travel to M.I.T. and frequent talk of the place for him to know that if he's being tested, this must be M.I.T.

Finally, as we stood out in the sun, the taxi arrived to take us across the Charles River. We bundled the awkward wheelchair in the trunk, him in the back seat, and we took the taxi over to Brigham and Women's Hospital where the magnetic resonance scanner awaited him. It is more than a device, it is the size of a room, and it is not pressed against the patient like an X-ray scope, but surrounds the patient as he is inserted *within* the machine.

Henry was a bit dubious about this, especially as he had to remove all metal objects from his person. That meant his belt, and he was shy removing it in front of the women researchers in the room. He was more chagrined after he removed it and everyone could see that there was a paper clip holding his trousers together.

I sat in the room with him, while everyone else retreated to another room, behind a large observation window. They spoke to Henry by microphone, trying to reassure him by tone of voice while they were in fact being distant and cold. And that noise! It was like being in a closet with a jackhammer operating in slow motion.

From all this crudeness issued, on the other end of the computing systems and wires, an elegant, flickering color image. A live image of his brain, slice by slice, as the imager moved backward through his brain taking images one plane at a time. When the imager reached Henry's temporal lobes and the place where his middle brain should be, the researcher at the panel let out a low gasp. "Oh! That's beautiful!" he said. Used to peering at subtle shading differences denoting massive tumors, he was now confronted with a huge black hole in the center of the brain, the first thing of its kind he had ever seen.

The series of images lead to measurements, recalculations, new guesses. There is, Dr. Corkin has discovered, a little more of the hippocampus present than was thought. But the other bits of middle brain linked to it, the parahippocampal gyrus, the entorhinal cortex, and the perirhinal cortex are all destroyed: they must be an important part of the "hippocampal" system of memory consolidation.

The pictures from within the dome of Henry's skull are a bit startling. Of course, we could have guessed what they might look like. But it is not the same to guess as to see the lovely textured images of the brain and then the black, ragged edges where tissue has been sucked out.

I sat in the uncomfortable surroundings at the hospital, an environment utterly unconducive to the sensitive discussion of thought and feeling, and it struck me as odd into whose hands we have given these mysteries. Discerning the plight of the human mind is a work left to the most extreme of thinkers: the scientists have access to the information which lies beneath the surface of things, but can neither effectively collect nor express what they may find there. And there are the poets, who have the gift to say what they think elegantly and convincingly, but know little of how things work beyond common cultural descriptions. It is as if the scientists are hunters searching the tundra's crystal land wearing heavy boots, while the poets fear to venture there at all, and so work only with the poor information that can be sent back in postcards from the frontier.

In the months during which I first thought of writing about the amnesic of M.I.T., my wife began to experience the first excruciating symptoms of her coming disease. These memories are neural assemblies of my own, reignited now as I remember.

As the dusk falls and the lights go on in the house, I sometimes think of a street in Cambridge—Quincy Street—where William James used to work. The lamps project pools of limpid light through which apparitions move. (It may well have been in this very room of the house that Dr. James wrote some of his lines about memory—that beyond the borders of the present moment "extends the immense region of conceived time, past and future." He said, "Into one direction or another of it, we mentally project. The stream of thought flows on; but most of its segments fall into the bottomless abyss of oblivion.") Why do some memories fall into black and other, more terrible, memories do not?

In spring of 1987, in Mary's last moments, I sat beside her, holding her cold hand. Through the clear tube I could see the liquid slide slowly down into her encrusted, discolored veins. The respirator could pump no more air into her one lung, as it was rapidly filling with fluid. Every few moments, her body would jerk, she would seem to gasp and struggle. Her most terrible apprehension, which had practiced upon her mind for three years, was that she would be in great fear at the end and be unable to communicate that. I asked for more morphine for her, and it was given. I waited, and one more time she stiffened. A noise jumped from her throat. In another twenty minutes she was gone. The doctor issued the assuring words; no, she could not feel anything, she could not know. I knew he lied; he did not know whether she could feel it or not. I still do not know. These doubts and images are now laid up among my neurons, and nowhere else.

Dr. James addresses me directly, his answer to the previous question in a footnote in his *Principles of Psychology:* "Who of us, alas, has not experienced a bitter and profound grief, the immense laceration caused by the death of some cherished fellowbeing?" he recalls Hodgson as saying. "In these great griefs, the present endures, neither for a minute, for an hour, nor for a day, but for weeks and months. The memory of the cruel moment will not efface itself from our consciousness. It disappears not, but remains living, present, coexisting with the multitude of other sensations." Having talked hours with Mr. M., I have briefly thought, for a moment anyway, that forgetting may be a blessed peace. He is himself protected by it from full knowledge of who he has become.

24. ☜

The plumbing of memory is most famously associated with Sigmund Freud. He was the first to devote himself to memory as both text and therapy. It was the very heart of analysis to wade into the dark waters, find what may be hidden there, draw it up for examination, and thus replace our frightened reactions with illuminated understanding. But Freud was impatient; he could not wait for science to discover how memory actually works. Freud himself said his own view of memory would eventually be replaced by facts issuing from neuroscience, and he was right.

To us, looking backward, the work of Freud has an antique quality, a confidence that an adventurous mind, loosed of its cultural and religious bonds, might well march through the jungle of neuroses and discover, all in one heroic expedition, the hidden temple at the center of humanity. As a neurologist, Freud was frustrated to be bound to the delicate and unanimated little masses before his microscope. He realized that if he were to remain a neurologist and continue to trace the physical evidence for mental action, in maddening labors month after month through his life, he might not arrive anywhere greatly more exciting than where he was at that moment. He preferred, as an avid climber of the Alps, to leave behind the painful work of scaffolding below and advance oblivious up the hill to a perch where all could be seen clearly and at once. The most remarkable summary of his work was penned by Freud himself in a letter to his friend Fliess in the winter of 1900. "You often estimate me too highly," he wrote. "For I am not really a man of science, not an observer, not an experimenter, and not a thinker. I am nothing but by temperament a Conquistador—an Adventurer, if you want to translate the word—with the curiosity, the boldness and the tenacity that belongs to that type of being. Such people are apt to be treasured if they succeed, if they really have discovered something; otherwise they are thrown aside. And that is not altogether unjust."

More than adventurer, though, he was a man who could tell the tales of his adventures with such urbanity and wisdom that it almost seemed unimportant whether he had got the details right. He was convincing, and psychology, art, and literature followed the adventures, enthralled. "One day," he wrote, "I discovered to my great astonishment that the view of dreams which came nearest to the truth was not the medical but the popular one, half-involved though it still was in superstition. For I had been led by fresh conclusions . . ." And so he carried us away.

There was some grumbling by those who attempted to pin down the details. Did the memory he imagined—a perfect record from birth, or even before, which could be consulted like an archive—actually exist? While he spoke of his method as scientific (despite his previous statement that he was no scientist), the motions did somehow seem like incantations as patient and doctor spoke, told tales to one another, and told tales about tales, until each felt better. Ironically, those critics who wanted to expose the mundanity and foolishness which seemed behind the method, found themselves not taken seriously. They ended spitting and sputtering before small gatherings, too evidently without the power to tell stories about behavior that were any better. Like too many scientists, they felt the facts should overcome the story. But both the facts and the power of the whole story are essential. One cannot abandon story merely because some of the facts don't cooperate. Freud's story about behavior, after all, was no brittle academic sketch. It was a vigorous venture into darkness, sexuality, and violence, which he said "lay underneath" the manners of daylight life.

One of the remarkable stories of Freud about memory is the one in which he read of a childhood memory of Goethe and determined to find its "meaning." Ulric Neisser included the story in a collection of pieces on everyday memory, noting "Why are a few, often apparently trivial memories of childhood preserved while so many others are lost? Freud, the thoroughgoing [nineteenth century–style] determinist, insisted there must be reason in every case."

Goethe wrote in his autobiography, "Across the street lived the three brothers von Ochenstein, sons of the late village mayor. They grew fond of me, and busied themselves with me and teased me in many ways. My family loved to tell all sorts of stories about the mischievous tricks to which those otherwise solemn and lonely men encouraged me. I will recount only one of the pranks here. There had been a pottery sale; not only had the kitchen been supplied for some time to come, but miniature crockery of the same sort had been bought for us children to play with.

"One fine afternoon when there was nothing doing in the house, I played with my dishes and pots in the rooms that fronted the street. Since this didn't come to much, I tossed a piece of crockery out into the street and was delighted by its cheerful crash. The brothers saw how much this amused me, I clapped my hands with delight, and they called out 'another!' I did not hesitate to fling the next pot and—encouraged by repeated shouts of 'another!'—a whole assortment of little dishes, saucepans, and jugs onto the pavement. My neighbors continued to signal their approval, and I was more than glad to amuse them. My supplies ran out, but they continued to shout 'Another!' I hurried straight to the kitchen and fetched the earthenware plates, which of course made an even jollier show as they broke."

Goethe tells us at the beginning that this was a family story, and not necessarily to be taken as truth. He goes on, "So I ran back and forth with one plate after another, as fast as I could get them down from the shelf, and when the brothers still claimed to be unsatisfied I hurled every bit of crockery within my reach to ruin in the same way. Only later did someone appear to thwart me and put a stop to it. The damage was done, and in return for all that broken crockery, there was at least a wonderful story, which amused the rogues who had been its prime movers till the end of their days."

With the gesture of a great conductor, Freud took this little tune, a lighthearted melody, and added drums and trumpets and great silences to draw from it much more than amusement. "In preanalytic times this passage would not have been disturbing and could be read without hesitation," he said. "But now the analytic conscience has come to life."

Freud quickly drew attention to facts of Goethe's life which fascinated him, and brought together elements he loved—the issue of early childhood memories, of sex and death: "It is not a meaningless or insignificant matter when some one particular of a child's life escapes the general forgetting of childhood. On the contrary, one must suppose that what has been retained in memory is also what was most significant for that period of life: either it already had great importance at the time or else it acquired that importance through the effect of later experiences. It is true that the great significance of such childhood memories is rarely obvious. Most of them appear unimportant or even trivial, and at first it seemed incomprehensible that just these could defy the amnesia of childhood. The individual who had preserved them through long years as his own personal memories could no more do them justice than the stranger to whom he related them."

The nineteenth-century certainty in the rationality of thought and behavior emerges here through Freud's tale. The mind must be a mechanism of wonderful wheels and gears— the gears which move the automatons of the clock tower, the spinning wheels in the brass, steam-driven machines were the strong metaphors of the time, and they must produce some balanced, satisfying end to their work.

Freud then seeks the source of this story or any which Goethe might tell of his childhood, by searching for what he felt must be the most significant feature of his childhood. In Freud's view, the purpose, and every gear in the mechanism, must somehow be related.

Freud then recalls a patient once reported that "around the time he tried to kill his hated brother, he had thrown all the crockery he could reach out of the window of his house into the street. The very same thing that Goethe describes in his personal recollections! But could the conditions necessary to support this interpretation be found in the poet's childhood?"

Goethe had not killed a brother or even hated one. But he did have brothers and sisters who failed to survive the insults of childhood. Four died between his second and his twelfth years. Freud found that according to Goethe's mother, Geothe, at the age of ten, did not weep upon the death of his younger brother Hermann, who was six.

Freud continues, "We can conclude that the throwing out of the crockery was a symbolic, or more precisely, a magical act. By this act the child [Goethe as well as my patient] gives vigorous expression to his wish that the disturbing intruder be eliminated. We do not need to deny the delight that the child takes in the crashing objects. The fact that an action is pleasurable in itself does not prevent—indeed, it invites—repetition in the service of other motives."

But why did the two stories both have their characters throw the crockery out the window, rather than just on the floor?

" 'Out' seems to be an essential component of the magical act, which stems from its hidden meaning. The new baby is to be removed through the window, perhaps because it came through the window in the first place." Freud here refers to his belief that all children were told that the stork brought babies in through the window.

Furthermore, Freud believes, "If we now return to Goethe's early memory, and put what we think we have learned from the cases of these other children in its place, we find a perfectly comprehensible connection that would otherwise have remained undiscovered."

The connection, perhaps somewhat less than obvious, is that according to Freud, Goethe did not want to share his mother's love with a new brother, and began his autobiography with this fact, disguised in a story the meaning of which even Goethe did not know. With a final flourish, Freud asked is it not perfectly appropriate for Goethe to mark the beginning of his work with a statement of his dependency on his mother's love? Freud was constructing a story within his memory of what might have occurred, what he emotionally would like to have seen as a satisfying tale.

But against this we must replace recorded facts, which disagree. There is more than one element in this tale that reminds me of the way Dr. Scoville handled his patients. There is in it a boldness and confidence married to a kind of insensitivity. Though Scoville disliked Freud's ideas, he was, like Freud, an adventurer. Each sought to solve a riddle with a single charge on the heart of the mystery, one with a knife and the other with a wit.

The mystery of memory is hidden in a tangle of neurons and will yield to neither instrument, but only to patient separation, thread by thread, of its soft machinery, which has accumulated through evolution like seaweed on the beach. Flinging a shoulder against the door of memory will not do. It takes a key—worse, many keys—to have any hope of entering.

I recently visited a psychiatrist friend of mine at the National Institutes of Health to ask him about memory and therapy, and what he thought of Freud and memory. It was clear that he had made the jump from Freud's thought only very reluctantly. He, Dr. Philip Gold, had been an English major in college, and he told me that before he considered medicine as a profession he had studied under Reynolds Price in college. He had written about the hero in literature, and had taken for his study the country priest in the writing of Martin Guzman. He got back a high mark on the paper and thought the note on it from Price read "Your writing has reached a symbolic complexity beyond which you should not go." On graduation day, he asked about the comment. "'I am afraid you have misread my handwriting,' he told me," Gold says. "'I actually wrote that your writing has reached a syntactic complexity beyond which you should not go.'"

Gold entered medicine, where syntactic complexity goes un-noticed. But he always focused on the outstanding figures, the intellectual therapists of great skill and compassion. They had, he said, developed ways of working with people to aid them in untangling the sometimes painful and maladaptive ways they had come to react to the world and to those they love.

"I only regret that the expectation, or the promise, or perhaps the fantasy that these techniques could be used to address de-finitively—to cure—terrible mental illness such as schizophre-nia, depression, manic-depressive illness, has not held true," Dr. Gold said. "There are other factors we have learned of, biologi-cal factors, which must be taken into account."

He had come to realize that talk therapy would never be sat-isfactory for people who are profoundly ill, and that even if some great therapists were able to make progress in intensive work with sick people, society could never mobilize the resources to use these methods with any but a few.

Eventually, Gold began the study of biological psychiatry and found as many satisfactions in the surprises of the brain as he had found in the struggles with the troubles of patients. He found that the hormones which are familiar to us have other lives, secret lives, in the brain. He found that much of the dam-age of mental illness could come from the physiological solu-tions set into motion to protect a person; these had caused damage by continuing to over-respond, causing more harm than help. He found that there are as many intellectual surprises, twists, and connections in the biology as in the relations with a patient.

For example, depression could be understood as a state of emotional hyper-arousal, he said. In it, alertness, tension, anxi-ety, and the reactions to events may become heightened. This is useful for short periods of time, in moments of extreme stress. But if the biological switches get stuck in this "on" state, great damage begins to accrue. First, chemically, the substances needed for a balanced and moderate way of responding become depleted. Later, the exaggerated reactions a person feels, and acts on, begin to cause damage to the love and trust between the depressed person and others. The strong and sometimes uncontrollably negative feelings lead to trouble and sometimes abuse. Children are alienated, marriages become rancorous.

Dr. Gold spoke of "monochromatic anger" created by biological depletions in the brain. This is anger that often appears inexplicably and is vented inappropriately on people nearby.

We spoke in a small white-walled office. The phone rang very loudly every few minutes, but Gold's compassionate tone was undisturbed.

In depression, there is often a strong, unremitting sense of the negative side of all things: "Sitting with a patient who felt this tremendous sense of anxiety . . . they didn't want to be who they were, didn't want to be where they were. They had no hope for the future. These are individuals who are biased toward interpreting everything about themselves—their world, and all the stimuli they receive—in a depressive tone. Negatively. They have the terribly lonely and vulnerable feeling that even their pain is a reflection of their deficiencies. If only they were more worthwhile and useful, they could adequately come to terms with their lives, then they wouldn't be so miserable, they would have friends, they would be successful.

"It is a terrible, vicious cycle. It did seem amenable to investigation and conversation . . ."

He gazed at the floor for a moment; I felt he must be remembering the difficult moments, the patients with whom he had failed. He would not speak of them, though. He looked up.

"But when we began to try new antidepressant medications"—he described the control groups, the placebos, the aspects which made it all convincing to him—"in about two or three weeks, almost predictably, patients with a certain kind of pattern—people who had depressive attitudes, inappropriate guilt, and so on, and who also had disturbed sleep or trouble eating, or diminished interest in sexual matters—would respond remarkably.

"That negative tone, that terrible anxiety, in about two or three weeks would soften up, and begin to lighten, even without the development of an alliance, without my knowing as much as I should know about their suffering," he said.

"That was a rather stunning and disturbing experience. To

see that the medication itself could make that difference. The idea that small biological changes could interact directly with their life experiences was strange," he said. "Without my performing any feats of virtuoso interpretation, or heroic feats of sitting with a person who was almost in intractable pain, suddenly the work was becoming easier."

Dr. Gold spoke of the curious sense of disappointment he had when a patient who talked for some weeks about anxiety, love, and guilt, and how they were twined with feelings toward other people, then, without warning at first, began to say, "Well, you know, doctor, I don't feel that way anymore . . ." The effect of drugs seemed to mock his own powers.

Shifting to an intellectual summary, Dr. Gold said, "The way I have come to conceptualize that now is that depressive illness can be characterized, in very loose terms, as a problem of memory. That is, when an individual has depression, they have preferential access to negatively charged memories. It is almost impossible for them to describe a time when they remember feeling well, or even to describe what it's like to feel well." The patients also found that when they are depressed, many of their experiences gain negative charges even as they are being laid down as memory, even if they in fact are neutral or positive. They are false memories of their own kind.

Gold said, "Sometimes people seem only able to remember what happened to them when they were depressed; they can remember where they put their keys when they were depressed, what they did and whom they talked to . . . Antidepressants can lift the veil of negativity and seem to unlock the capacity to have access to the whole repertoire of memories and their feelings."

It is ironic, he said, because in his training some years ago there was a feeling that medications were poison but that they might cover the intrinsic self of the individual. In fact, with new antidepressant drugs that affect more precisely those arrays of neurons involved in negative feeling, matters have resulted in the reverse: "The antidepressants actually liberated the individual to be able to discuss more readily the fears and fantasies about himself, the level of anger or rage or disaffection with others ... [They made it] possible to reconstruct stories of your life, who you are without the wrenchingly negative associations and without a flood of completely negative memories," he said.

The drugs have allowed us "to do what otherwise would be inconceivable, to rethink one's life, in the light of information from memory one otherwise didn't have. We can come to realize that in depression one had spent one's life almost inauthentically."

These things are not widely known, Gold said. He worries that the prejudice against drugs may still hinder the understanding of what has happened in recent research, or that the uses of conversation will become too easily ignored by others who have become enthralled with the drugs. He has some hope, though, that issues of brain and memory will in time be sorted out.

25. ❧

As I thought about it later I returned to the point that Gold had made about the drugs interacting directly with the patient's life experiences. The brain, as we know, is physically altered almost constantly as we experience the world. The input of the senses alter our brain chemistry. George Johnson, a colleague and the author of *In the Palaces of Memory*, wrote at the beginning of that book about his astonishment at this seemingly obvious fact: "Whenever you read a book or have a conversation, the experience causes physical changes in your brain. In a matter of seconds, new circuits are formed, memories that can change forever the way you think about the world. . . . It's a little frightening to think that every time you walk away from an encounter, your brain has been altered."

In such a fast-changing biochemical system, the idea of a fixed "personality" thus may be a problem. Our images of ourselves are themselves changeable, as we now know from Henry M., Kent C., and others whose personalities have undergone massive alteration. They have found new equilibria, new stories about who and what they are.

We have learned since Henry's advent that the constructive

nature of memory may lead us to dangerous assumptions, both in therapy and in other social institutions which include some assumptions about how the mind, the memory, works.

It is only now understood that memory and imagination are at base the same process, and they can contaminate one another. Because of the abstracted nature of our grasp of the world, we lose directness, but at the same time we gain another power. In the world itself, a rock may not bend, nor may it merge with rocks of other colors. But the representations of these among our neurons can bend or mix in any sort of chimera, any felicitous or fearful hybrid. Neurons react not only to direct sensations from the world but to signals from each other. What are uncooperative solids outside us may be combined and recombined as easily as colored dyes in a transparent gel, as easily as shadow and light merge.

And they accumulate! Both the particulars of life we remember, and the abilities we earn from teaching each other across the generations, accumulate. Both the real and the chimeric may become either the lucid symbols of mathematics or the false feelings of madness.

One scientist who has studied the subject, and who has since become a well-known witness for the defense because of it, is Dr. Elizabeth Loftus of the University of Washington. I called her to talk about the reliability of memory when I was investigating a case of murder in North Carolina, a case in which it appeared that four young men were convicted based on the "enhanced memories" of one suspect, who knew nothing of the crime until he was hypnotized. He then confessed falsely, led on by the hypnotist who fed him details of the crime, which then were integrated into his "memory." Fortunately it was possible to review a videotape of the hypnotic sessions and identify the moment at which the psychologist introduced each fact about the crime that the young man did not know and to trace the growth of that idea in the young man's tales as the psychologist pressed him for more fake details.

Dr. Loftus began in the 1970s to explore the nature of forgetting, the beginning of a wave of new work on memory in the practical situations of police precincts, court rooms, and psychologists' offices. She was curious about what people believed about it. People seemed to believe both that forgetting occurs and that memories are permanent records which, even if we cannot recall things sometimes, are still available to be recalled if we could only find the way—though therapy, or hypnosis, for example. She published a survey in which psychologists and laymen were asked whether they believe that "everything we learn is permanently stored in the mind, although some details are not accessible" or the converse, "some details that we learn may be permanently lost from memory . . . and would never be able to be recovered by hypnosis or any other special technique." Of the laymen, 69 percent agreed with the first proposition, while 84 percent of the psychologists agreed with that proposition— odd in some ways, because the "tape recorder memory" theory was disproved decades ago, and one would have thought psychologists would have a more recent and complete grasp of the scientific data on the subject than laymen.

In fact, when Dr. Loftus asked the psychologists the reason for their belief, many cited the work of Wilder Penfield and his studies during lobotomies and other brain surgery. It was later proved that they were not memories, as Penfield had said, but simply fantasies, more like dreams than memories, which included many impossible facts or scenes for which the patient could not have been present. Perhaps the psychologists' memories of the debunking of the Penfield speculations, which occurred by 1970, was overpowered by their wish that the work was true.

Loftus began to devise experiments to test the idea of the permanence and accuracy of memory. Beginning with the notion that the ideal circumstance would be to listen to one person relate the memory of a real event—say a traffic accident—while someone else reviewed a videotape of the actual event, what would the differences be between the memory and the tape?

In an experiment which is now among the most important in the recent history of psychology, Loftus showed volunteers a series of slides (videotape was considerably rarer in the 1970s, when the experiment was done) in which a car, a red Datsun,

moved along a street to a traffic sign, then turned right, knock-
ing down a person crossing the street in the crosswalk. Half the
volunteers saw slides in which the traffic sign was a stop sign,
and half saw a yield sign.

People remembered the details of the accident reasonably
well. Most could say what kind of traffic sign they saw. But Loftus introduced a difficulty. In one series of tests, some of those
who saw a yield sign were asked on a questionnaire, "Did another car pass the red Datsun while it was stopped at the stop
sign?" The misleading information, even though it was only
mentioned and never reinforced with information from the researchers, quickly overrode what the volunteers had actually
seen. They remembered what they were told in passing they recalled, not what they actually saw. Worse, when the volunteers
were probed about whether they were sure of what kind of traffic sign they saw, they replied that they were sure. When shown
two slides side by side, the one they actually saw and the one
the question later suggested, the volunteers most often picked
the one suggested, not the one seen, and were certain they were
right.

Loftus continued with similar experiments involving different scenes and different ways of suggesting material, with different ways of presenting memory. Her conclusions were
disturbing. People were extremely suggestible. She found that
eyewitnesses are often wrong about details and sometimes
wrong about the most startling points—for example, the race or
gender of someone they saw in an experimental "witnessing of a
crime." Those witnessing a crime often will find later they inad-
vertently alter the criminal's race to fit their own initial prejudices about who is more likely to be a criminal.

In later experiments by others on hypnosis, it became clear
that people under hypnosis were even more suggestible. When
asked to remember every detail and "zoom in" on some they
were unlikely to have seen at all, they obliged. When the experimenter did not give a hint about what kind of detail to come up
with, the subjects made up details randomly. When the experimenter gave hints or leading questions, the subjects obliged
with false memories to fit the expectations of the questioner.

The effects can be completely unconscious and extraordinarily subtle. In another experiment, people were shown a film of an accident and asked questions about it, such as "About how fast were the cars going when they smashed into one another?" Others who saw the film were asked the same question, but the word smashed was changed to hit. Later, both were asked if they remembered seeing broken glass in the film. There was none, but those who heard "smashed" also reported later that they remembered seeing broken glass.

Loftus and her successors in studies of memory and forgetting have learned that it is most likely that much of what a person experiences is lost permanently, but in any case can be easily and irretrievably altered as soon as someone asks about it. In rather understated prose, she concludes that the implication of the notion of memory that is nonpermanent "is that it should give pause to all who rely on obtaining a 'truthful' version of an event from someone who experienced that event ... Clinical psychologists, counselors, and psychiatrists ... Anthropologists, sociologists, and some experimental psychologists query people about their past in the course of studying some particular problem ...

"The contents of an interview may not reflect a person's earlier experiences and attitudes so much as his or her current picture of the past. It may not be possible, in some instances, ever to discover from interviewing someone what actually happened in that person's past. Not only might the originally acquired memory have departed from reality in some systematic way, but the memory may have been continually subjected to change after it was intially stored."

She wrote these things in 1980. The discovery of the fallibility of memory and the ease with which psychiatrists or policemen or reporters can influence memory began in the 1970s and has continued to become stronger and more clear. Other notions about memory have also been radically changed with research—for example, that childhood memories are important to present psychological states. Childhood memories, in fact, don't really exist at all until after age three or four, and are for some time after that quite sketchy.

Further studies have shown that half or more of those who are later found to have been unjustly convicted of a crime were convicted on the basis of erroneous eyewitness testimony. Perhaps half of one percent of eyewitness testimony is wrong, and if so, according to these estimates, five thousand to ten thousand people a year are jailed because of false memories.

I discovered the disastrous effects of false memories for myself in rural North Carolina, in the case which brought me to Dr. Loftus.

It was a delight to drive up to Union Mills, a small town in the piney hills of western North Carolina, but I drove there to investigate a murder case, the worst crime in memory in that area. Sheriff Damon Huskey talked to me; I was a reporter at the time for the *Washington Post*, a couple hundred miles and a couple layers of social rank north of Union Mills, and he wanted to convince me. He described the crime in detail. Eighty-eight-year-old Nannie S. Newsome, crippled by arthritis, a thin and frail woman who had taught school in the town for decades, had been dragged out of her tiny white clapboard home. She was found in the ballfield behind her home the next day, a January morning, beaten, raped, and strangled. She was in bloody pajamas, and lay over a garden rake; her face, knees, and feet were bloody. She had four broken ribs and strangulation bruises around her neck.

It was an odd crime, something beyond regular imagination, especially in that warm, relaxed piece of geography by the lower end of the Blue Ridge mountains. Even stranger than the crime, though, were the initial suspects, four young black men who did not know the frail woman more than they did anyone else in town and had no apparent motive to attack her. Nothing appeared to be missing from her house, though the police could not think of a motive other than robbery. There were footprints in the house and the field, fingerprints, clothing fibers, and hair.

It appeared at first no one would be charged with the crime, but people were angry. Eventually a nineteen-year-old neighbor of the woman, Reece Forney, who had mowed her lawn, a man whose intelligence by I.Q. tests was 74, in the range of retarded, became a suspect. He was questioned, given a lie test with a voice stress detector, and hypnotized, all without the aid of a lawyer.

It was the "enhanced memory" of Reece Forney, drawn out under hypnosis, which created the case against him, his brother, and two friends. The psychologist who hypnotized him, strictly according to the rules of such encounters set out by police and hypnotists, told Reece that he would be able to see the events of the night of the crime, could move things forward or backward in memory, and zoom in on details he did not know before.

Forney said he passed the house that night and saw a shadow in the window of Miss Newsome's house. When he "zoomed in" it turned out to be his friend Richard Flack. It was a dark night, he was 125 feet away, and could not by human vision have detected anything more than a shadow. Nevertheless, the story went on: Reece went home to bed and was awakened by Lester Flack, another friend and brother of the man he had placed in the window. Lester, he said, picked him up from bed, threw him over his shoulder, jogged a mile to the house of another friend, Stephen Hunt, and the group, with Reece still being carried, ran to the house of Miss Newsome, where Lester picked her up, threw her over the other shoulder, and ran outside to commit the crime with pals in tow.

The physical evidence—foot- and fingerprints, hair, clothing fibers—matched none of those whom Reece Forney had accused in his enhanced memories. The case turned on crucial evidence like the rake with which Miss Newsome was raped; it was not mentioned in newspaper accounts, and so Reece's recollection of it suggested he must have been there.

The videotaped memory-enhancing session, however, shows it was the psychologist, after being handed a note by a policeman, who first introduced the subject of a rake.

Forney: [describing walking home after the crime] Seems like I grabbed something and ran back . . . I walked most of the way because I was tired.
Psychologist: [handed a note by policeman] What did you grab?
Forney: Base of something. Base of something.

Psychologist: Was it a rake?
Forney: I don't know. It could have been.
Psychologist: Where did you get the rake from?
. . . .
Psychologist: And what did you do with the rake?
Forney: I don't know.

At each point, when Forney said he did not know, the psychologist suggested something . . . Did you have sex with her? . . . Now somebody is strangling her . . .

With each questioning and each new session of hypnosis, the tale changed. Four men were convicted of the crime, despite the fact that the physical evidence matched none of them.

There has been a recent wave of accusations, and convictions, based on what are now called "recovered memories" arising out of sessions of hypnosis or therapy. Research has showed that under hypnosis or even under extended questioning by police or therapists, the number of "memories" a patient can bring up increases greatly. Unfortunately, more than three-quarters of the retrieved memories are completely false and are often found to be suggested by the therapist.

There is no limit to the amount or strangeness of the material that can be "remembered" by patients who are open to suggestion. Worse, when asked later about memories that were planted and were intentionally false, the patients were firmly convinced they were real memories. Suggestion in its full variety of forms, from therapy to questioning, can alter memory, destroying whatever may be legitimately retrieved from an interview.

But this understanding is not widespread; police departments routinely use hypnosis and other forms of suggestion, therapists still routinely suggest memories to patients, and journalists routinely suggest interpretations to sources. There is little appreciation in our current behavior and institutional practices that suggestion alone may change memories.

The cases of "recovered memory" now number in the hundreds and approach the thousands. And there are other areas of life, more persistently with us, that remind us that the conviction of memory is not infallible. Recently there has been a variety of testimonials from people claiming to have been abducted by aliens in flying saucers, and even a sober article in the *Washington Post* by a professor from Cambridge Hospital, a Harvard affiliate, that says we should believe these people because they truly remember and believe in these encounters.

A movement has quickly grown up around the idea that a person can have no memory whatsoever for a traumatic event of childhood, then, twenty or thirty years later, "recover" it in detail. There is nothing in neuroscience which suggests that such a thing is possible, and there are no cases in which such phenomena have been demonstrated. In fact, the movement depends not on the memories of the patients but in every case on the insistence of the therapist. As one "recovery" therapist, Renee Frederickson, writes, there can be no doubt about what the therapists suggest. "You may be convinced that your disbelief is a rational questioning of the reality versus unreality of your memories, but it is partially a misguided attempt to repress the memories again." This sort of reasoning, or failure to reason, seems fantastical, like a throwback to prescientific times. Frederick Crews, a critic of Freudian thought, points out that the source of this kind of belief—belief in a memory that retains everything for all time, and a memory that can completely forget then recover important events later—is early Freud. Crew quotes from an article which discusses confabulation of memories that Freud may have created in his patients, and from Freud himself: "The work keeps coming to a stop and they keep on maintaining that this time nothing has occurred to them. We must not believe what they say, we must always assume, and tell them too, that they have kept something back. . . . We must insist on this, we must repeat the pressure and represent ourselves as infallible, till at last we are really told something. . . . There are cases, too, in which the patient tries to disown [the memory] even after its return. 'Something has occurred to me now, but you have obviously put it into my head.' . . . In all such cases I remain unshakably firm. I . . . explain to the patient that [these distractions] are only forms of his resistance and pretexts raised by it against reproducing this particular memory, which we must recognize in spite of all this." The repressed memory move-

ment, Crews shows, is a throwback to adventurous and sugges-
tion-aggressive Freud of the last century, for whom there was no
attempt to verify anything. He simply believed in the fierce spirits he perceived moving behind his patients' mundane tales,
and insisted on them.

It has now been demonstrated that memories are actually
rather fragile, easily altered and easily created. In cases of
trauma, the problem is never remembering terrible events but
always trying to forget or diminish them, in order to carry on
with life.

"The 'drawers' holding our memories are obviously extremely
crowded and densely packed," Dr. Elizabeth Loftus writes.
"They are also constantly being emptied out, scattered about,
and then stuffed back into place.... As new bits and pieces of
information are added into long-term memory, the old memories are removed, replaced, crumpled up, or shoved into corners.
Little details are added, confusing or extraneous elements are
deleted, and a coherent construction of the facts is gradually
created that may bear little resemblance to the original
event.... Every time we recall an event, we must reconstruct
the memory, and with each recollection the memory may be
changed—colored by succeeding events, other people's recollections or suggestions, increased understanding, or a new context.
Truth and reality, when seen through the filter of our memories,
are not objective facts but subjective, interpretive realities. We
interpret the past, correcting ourselves, adding bits and pieces,
deleting uncomplimentary or disturbing recollections ... thus
our representation of the past takes on a living, shifting reality."

Jean Piaget has recalled in his writings several false memories
which he discovered later. Among them is this remarkable one:

"One of my first memories would date, if it were true, from
my second year. I can still see, mostly clearly, the following scene,
in which I believed until I was about fifteen. I was sitting in my
pram, which my nurse was pushing down the Champs Élysées,
when a man tried to kidnap me. I was held in by the strap fas-
tened around me while my nurse bravely tried to stand between
me and the thief. She received various scratches, and I can still
see vaguely those on her face. Then a crowd gathered, a police-

man with a short cloak and a white baton came up, and the man took to his heels. I can still see the whole scene, and even can place it near the tube station. When I was about fifteen, my parents received a letter from my former nurse saying that she had been converted to the Salvation Army. She wanted to confess her past faults, and in particular to return the watch she had been given as a reward on this occasion. She had made up the whole story, faking the scratches. I, therefore, must have heard, as a child, the account of this story, which my parents believed, and projected into the past in the form of a visual memory."

All of these mental shenanigans occur in the traffic between the hippocampus and the dome of the brain, which we understand only in the sketchiest manner scientifically. What is most curious is that the rush of new information about memory and its fallibility has occurred since the 1970s. This is the same period during which society became afflicted with hundreds of cases in which memories of crimes, satanic rituals, and child abuse were recalled decades after the event only from the coaxing of a therapist. It is a sign of disjuncture between the culture of science and the practical world of crime and therapy that now, at just the moment when we should understand that these memories can be completely false and still provide the rememberer with complete confidence that they are true, we nonetheless have great difficulty acting on that knowledge in psychiatry, police work, reporting, or in court. It may be no accident at times.

Robert Jay Lifton studied the memories of Nazi doctors who helped out in the concentration camps during the Second World War. How is it possible for these men not only to live with themselves, but to continue to have rather high opinions of themselves? He concluded that they use a form of self-delusion which he calls "middle knowledge" —a form of knowing and not-knowing at the same time. One doctor, Lifton reports, was involved in the shipment of large amounts of cyanide to the camps. He remembered his involvement, but, at the same time, was shocked when he was told directly that the gas he sent was used to kill large numbers of people. He had worked very hard on those memories, and eventually he had rendered them harmless enough to get past them. And so it is with those who experience childhood abuse or trauma; the difficulty is not that they don't remember it. They do remember, in detail; but in order to heal, they must learn how to distance it. Our memories must not be too perfect, after all, but must allow us to carry on. (To what degree does H.M. set aside whatever he might remember of the surgery which hurt him forty years ago?)

It has only begun to be clear in recent years that trying to bring back trauma, as is automatic in some talk therapies, and to recall the details of the trauma, may sometimes be the opposite of what is needed for mental health. The brain naturally submerges, forgets, and retells the stories of the past to accommodate our current view of ourselves. Forgetting bad experiences is important to healing, as is retelling them to ourselves so they seem less threatening.

26. ⤞

It has been almost ten years since I began to seek out Henry
M. and memory. In the meantime, my wife has died, and I
buried her to the words "Out of the low door they stoop into
the honeyed corridor, then walk straight through the wall of the
dark." My wife has become a memory. My children have grown
and their small charms are memories. And I have met Henry,
and he has become a memory as well. Now when I walk in
Cambridge the air has a bittersweet edge. I feel differently
about mind and memory, having understood what it means to
lose it, and having begun to understand the mechanics of having it.

Our understanding of memory has shifted rapidly since the
time that Henry M. first entered the Hartford Hospital and Dr.
Scoville began to treat him. The insights of Freud have given
way; drugs have become useful in treatment.

We are still ignorant but have learned some of the rules by
which neurons operate. The path from sensation to meaning,
for example, from the first glimmer of a face to the flush of
recognition that it is Grandmother, has been traced in the brain.
It has become clear that the mind is manifold, made in parts

and working as parts on many computations simultaneously, bundling them together only as needed. It has been shown just how delicate memory is, how fickle and changeable, as it forms and reforms after the fact. It has become clear that the act of memory is an act of construction, not of recording. That is, we create experience and memory as we go, rather than being mere registers of the events around us. We reshape memory as we move through experience, and reshape experience with expectations brought up from memory.

What we leave behind as we learn more is the fixed model of memory and its attendant fallacies. The brain systems which record episodes of life do not even begin operating until the age of three or four, and then only haphazardly for some time. The power of memory grows with the power of language, and memories must be able to fit with other knowledge in order to be laid down.

In reflection upon these things, after a time, I began to feel that the heart of memory's mystery is not actually the memory. It is the act of experience itself that is most mysterious.

We do not actually see the color of objects; we merely pick up with limited antennae—the rods and cones—a few of the different wave forms, in a narrow range of vibration, among the many deflected-off objects. Because we sense them as different, we have named them "colors." But the colors are not properties of objects; they occur between the object and the eye. We do not hear the sounds of the world either; we merely pick up a few of the scattering waves of pressure in the air, in a narrow range of frequency, as they tap against the stretched skin drums within our ears, and we have named them sounds. We do not smell the fragrances of the world, but merely pick up a few of the escaping chemicals from the surface of nearby objects, in a narrow range of shapes that well fit the detectors within our noses, and we have called them smells.

We absorb only a small part of the richness in the world, some of which is there for other animals, and some may be experienced by none at all, ever. I recall my surprise when I read that bees' eyes can pick up much of what ours can, but in addition they can see a wavelength of light, a color, impossible for us to perceive.

As a bee glides down toward what appears to us a pure white flower, it can see great dark crosses guiding it to pollen; these patterns exist in the world (in a color now called by entomologists "bee purple" because it is somewhere off the purple end of our vision), but we will never experience it. With our profundity of tricks, we can learn of it, because we have found film which can register it, then we may print the film in black and white, to see the markings. But we will never do better than seeing this color in black and white, because we simply haven't the equipment.

Psychologist Robert Ornstein reminds us that many people are startled to discover that our perceptions are not, in any large sense, true. He says, complete with the rampant italics of the emotive 1960s and early '70s, "There is no color in nature, no sounds, no tastes. It is a cold, quiet, colorless affair outside us. It is we who create colors from similar, though shorter, vibrations, and it is *we* who transform molecules that happen to fit into spaces in our tongue into steak and *sauce bernaise*—these things are *dimensions of the human experience*, not dimensions of the world outside ... We don't actually experience the outside world—we grab only a very refined portion of it, *a portion selected for the purpose of survival.* This human selection of reality keeps us out of trouble, it allows us enough information to run the body, to keep healthy, and most importantly to reproduce and survive. The mind is designed, not as we like to think, for thought, for creativity, for appreciating the opera, but rather to *allow us to respond to the immediate contingencies of the world outside.*"

Susan Allport, a biologist who visited the Marine Biological Laboratory in Woods Hole, Massachusetts, recalled conversations there about the gap between what we experience and what is in the world. Neurobiologists now revel in the awareness that our conception of the world is a mere facsimile of what's really out there, an abstracted and greatly reduced reality that has somehow been transformed into the language of the brain—the action potential, or firing of the neuron, in its characteristic assemblies. She asked Rodolpho Llinas, the short, wide-eyed and eminent neurophysiologist from Colombia, "So what is out there?"

He said he did not know: "I know there is no light out there. I know there is no color out there. I know there is no sound or taste. But there is something out there, something which I translate and make into a model of something I like."

Allport wrote that it was a beautiful blue-skied day, and Llinas and she were sitting on a patch of grass overlooking Eel Pond, the small tidal pond around which the village of Woods Hole is built: "'Do you ever forget that it's so?' I asked Llinas, who once told me that he left his very comfortable and affluent life in Colombia when it became obvious that the cook (he was age 14) could not tell him what he wanted to know about synaptic transmission. 'No,' he answered quickly. 'It's part of me. It's so beautiful. It makes life so unbelievable . . . what is really amazing is that the brain can do this and do it so well that we are convinced that what we perceive and what is out there are one and the same thing.'"

The experience within us is not of the world itself. And our memory is not a memory of events themselves. Both are only the feel of neurons alight. We are held apart from the world by at least the distance across our skin and must live inside ourselves. Our perceptions, memories, and the thoughts which arise out of their interaction are reflected and re-reflected among our mental mirrors. We discern persons and things about us, in Joseph Brodsky's phrase, like "Illuminated animals, showing through the black oilcloth of the night water's surface."

This, then, is the secret of memory: that it is not memory of the world at all, but a re-experiencing of the same neural chords in recalling an event as when we first saw, felt, or heard it. The mesh of neurons set fluttering in response to the sight of a tall palm tree as it moves in warm wind can be repeated, even though I no longer stand there beside it. Do I "see" the slender brown arc of the tree again? I do set alight an assembly of neurons, some of which were the same or in the same pattern as when I first saw the lovely shape. But I cannot bring back the thing entire. Some elements drop away, or are summarized, so my memory of a Belizean evening is a paler, but still welcome, version of the original moment. And of course, the memory in time may become contaminated by other memory or mere wishes. Did I actually see that phosphorescent fish there, or was that a fragment from another holiday trip? Or even from my readings about marine species which glow? And who was with me? Was it you?

Each time an assembly of cells is lit again, seeing the beach or just remembering it, the connections among its members, the parts of the scene, are strengthened and become more likely to be able to be fired up as a group, that is, *remembered*, when any element among its members is stimulated.

There are those who object to this description. British philosopher Colin McGinn has written that "somehow, we feel, the water of the physical brain is turned into the wine of consciousness, but we draw a total blank on the nature of this conversion. Neural transmissions just seem like the wrong kind of materials with which to bring consciousness into the world." This is a poetic objection, and he concludes poetically that the problem of understanding consciousness is most likely unsolvable. I think his feeling is like the one physicists had when they first comprehended the Einsteinian universe. Some refused to believe; Einstein himself had doubts. If all is relative, where can we lean for support? Within memory, if all moves, what is still? Memory is not certain or stable, but moves, even as we query it.

In day-to-day life we don't worry, though. We speak to one another as if we have a complete and satisfying rendition of the world outside us. From our small horde of sensations, we have the illusion of wholeness; we feel that we sense the world entire. This disparity between experience and what we now know must exist, even though we have just now begun to understand it, is something we have not yet assimilated, not yet explored in philosophy or social science to find its ramified meanings. I suppose it takes decades, perhaps centuries, for a culture to establish a radically different way of "seeing," and so it must be with this new understanding of the mind and the manner in which it operates, an understanding we now approach for the first time. It will require the adaption of a new perspective on life and society, as we are able to absorb it more fully. For the present, we know enough facts to recognize that some of the ways we act and some of the rules of our institutions, which are based on misapprehensions of how minds work, are erroneous. They are misperceptions analogous to optical illusions in the way we view behavior, analogous to our blindness to bee purple.

And so it is, after some years of exploration and thought, that we must conclude that the central feature of memory is its malleability. It is changeable upon the instant. New information adds to, overlays, or confuses old feelings, thoughts, and knowledge. Memory is, at the end, a site of endless construction where facades come down, beams are shifted, walls are sucked together or blown apart, all in response to the current, most urgent needs.

The recall of a crime may be one thing at the time of the crime, but when detectives loom above you with certainties of their own, and questions which lead into the path they want you to walk, then memories may move and rearrange themselves around a new need.

It is this malleability which philosophers, scientists, psychologists, judges, and historians, not to mention journalists, have worked to deny over the centuries. Their work after all is made so much more difficult if human memory does not contain certifiable facts, like rocks, upon which they can build. Nevertheless, we now see, memory is not solid; it is liquid.

Science is demonstrating this fact, and soon we must cope with it historically, socially, legally, psychologically.

Mental life may be imagined as a continuous storytelling—taking bits and fitting them into a running narrative that makes sense of where we have been, what's going on now, and what to do next. It is stories that make some kind of sense of the welter of data from outside, where there is no sense. It is we who must make sense of things. "Stories" are merely the structure we make with selected bits of input. We meet two Italians and so, on meeting the third, expect something and are pleased to find it. Now we have some kind of internal rule about Italians. True, it could be overturned on further experience, but we use our expectations like pitons to scale the side of life.

The central engines of our mind are bent always and forever on the job of making stories, in large themes and a thousand subthemes simultaneously. The brain's operators within the frontal lobes work together with memory on this.

Because it is what we do, and we cannot fail to do it, it is the fundamental bias of the mind. Not ideological or temperamental, it is a biological bias.

It is manifest in, for example, our understanding of death. It carries such import because it represents the sudden collapse of a great mass of knowledge and expectation, that which we know and feel and want about someone who occupies some substantial part of our whole consciousness. We imagine our beloved, and cannot project her or him to become nothingness. Our expectations run on, despite death, and then find we are mistaken, she is gone. Where there was some light, there is only a blank. Nothing goes forward; all must be revised. A great hole is opened up in our thinking and feeling of the world, a piece of our map of the world is gone. It is, with those close to us, a cognitive as well as emotional crisis that is not easily accommodated.

In this way we can see that the belief in an afterlife is almost a necessity of human mentality. It is a deep, biological tendency. We know only life, going forward in the world, the constant run of experience within our own consciousness, and of watching others in the world, and of history itself—all continue forward, all require knowing and expecting and reacting. Every fiber of our thought leans toward the expectation that we will go on. Death is the sudden breakage of all our long-built-up knowledge that things go on. We cannot picture it; we are incapable of imagining that we are not, that we are not experiencing. By nature it is an impossibility. The mind, the machinery of expectation, has no purchase there. Death violates thinking, so our expectations run on into what "it must be like" to be dead. Our expectations shoot out into the void without us, and we canot help but imagine "life" after death. We imagine our bodies going on, but at the same time know they do not, so we imagine some peculiar spirit-replicas of them going on for us. We cannot help it; it is a conclusion that springs directly from the mind's own machinery.

This is also one description of the origin of religion: it is the story that cannot stop. It is the tale of ourselves when we are not. How could the tale of ourselves simply stop? Our imaginations fail; the mind supplies stories to avoid the shock of the brutish blank that is the world in itself. Science, itself like a religion in that way, perhaps the religion of the modern mind, attempts to reapply our thinking to come up with other stories that better fit the facts, or do a better job fulfilling expectations, creating workable "scripts" we can live with.

Grief, as I know, is in some way a product of memory running on by itself over the edge and into the blackness. Our senses tell us that we no longer have the ground beneath our feet; we have overshot the cliff and are about to fall. Grief is this fall from light and life, into the dark of what is not known or expected. Grief is fear.

Our construction of the world by memory and expectation is a central engine of culture. "Culture is no more and no less than shared expectations," says Roger Schank. It must be taken into account, its natural prejudices and tendencies, its inability to understand easily some things and its all-too-easy ability to expect others, must be brought within our thinking about society and how we behave if we are to get a grip on the deepest matters—violence, deceit, honor, belief, and goodness.

Genes and physiology, of course, set the limits and prescribe whether it is hammer or sling we work with. But the new understanding of memory's importance tells us that experience may be the difference between mundanity and genius. To nature and nuture, we must add a third term.

This has been verified often in recent science, demonstrated, for example, in a series of experiments by researchers at Carnegie-Mellon University, led by Herbert Simon. They have used a number of challenges. For example, it was well known that chess masters could glance at a chessboard for five seconds or less and then reconstruct the position of every piece, as well as analyze the state of affairs between players. But Dr. Simon showed some time ago that such a feat was limited to arrangements found during actual games. Chess masters did no better than college students ignorant of chess when it came to memorizing the positions of chess pieces placed randomly across the board rather than in the orderly, cohesive patterns of a chess game. Simon's group showed that a student ignorant of chess could, with intense practice over fifty hours, duplicate the feat of a chess master.

Dr. Anders Ericsson, who worked from Florida State with the Carnegie group, also showed that students could listen to a string of random digits pronounced, and with practice at the art, could repeat strings of as many as 20 digits in a row after a single hearing of the list. One student was able to reproduce 102 digits without error on a single hearing after practice in the technique.

This is the reason why the greatest performers in many disciplines, from the violin to the 10-meter diving board, begin as children. It is the training and great elaboration of memory that is at the core of prodigious performance. As Dr. Simon says, "It can take ten years of practice to excel at anything. Mozart was four when he started composing, but his world-class music started when he was seventeen."

"The mind is so full of expectations that it is a kind of prediction machine, assuming what will happen next in every aspect of the world," writes Roger Schank. "Thinking, especially conscious thinking, means checking out why our expectations have failed. For the most part, we find ready answers. We can find other times in our memories when these expectations have failed, and we can compare the situations. We learn from experience by altering our expectations according to experience."

Again, memory is not an object but an act of recollection, involving the way we first experience an event, the many and varied links we have made to this or that experience in the time since we had it, and finally the way we attempt to search for it again.

We must imagine memory as an act: fishing with the right rod, with the lure that seems best, in the stream of choice. We may bring up a trout, and thus revive the moments in which we have landed other trout, and perhaps the dinners that followed. Or remembering may be imagined as batting, the response to a pitch, which is an array of motions and thoughts, changing even as the pitch drops.

So, we must ask, when a witness comes to court, what parts of a memory are ready to be activated for the jury? Are they the ones rehearsed by counsel or by police? Are they the ones that may not even be within memory, but instead within possibility or familiarity? We cannot know unless we make some effort not to contaminate the memory of the eyewitness, which is exactly what does not occur in our current understanding of witnesses and their memories.

For example, it is important to create the gist of a story soon in order to remember it. It is important to tell it to someone soon, or at least rehearse ourselves. If we don't retell it soon enough, the experience, as Roger Schank says, "cannot be coalesced into a gist." That is, "while parts of the experience may be remembered in terms of the memory structure that was activated—a restaurant may be recalled through cues having to do with food, a place, or the particular company—the story itself does not exist as an entity in memory. Any generalizations that might pertain to the whole of the experience would get lost. We could remember the restaurant, but we might forget that the entire trip had been a bad idea . . . the opposite side of the coin is also true. We fail to create stories in order to forget them. When something unpleasant happens to us, we often say, 'I'd rather not talk about it,' because not talking makes it easier to forget."

And now, in the wake of hundreds of false memory incidents, we know that telling a story over and over with the inclusion of details known to be untrue can produce a story which is completely false, from which inferred—and also false—details may be created. They may seem to be part of memory because they are a part of the story construction which we normally use and rely on in memory.

In order to know what is essential and what is not essential within an orderly society, we must understand these issues of memory, expectation, and behavior. Does pornography lead to violent crime or not? What laws are needed, and what are useless? The answer lies within memory and expectation, as well as in jurisprudence. How can doctors, lawyers, and journalists best be governed so they will police their own misbehavior? How much outside interference is necessary? Fundamental understanding in memory and expectation is also an important element in understanding these issues. Not that we could not come up with the right answers to these questions, in fact, we probably already have the right answers written down somewhere. But those with the right answers and those with the wrong answers will argue endlessly, until we have convincing evidence from within our own nature as to which is the right path.

By including these matters, we will be adding the third dimension to social thinking. I suspect that to understand these things we must soon abandon our idea that the genes lay out the plans and possibilities, and the environment dictates what will come of it. What is missing is the dimension of action. It is the missing element, representing the complexities which we don't understand. They are, after all, more complex than either genes or environment. Genes and environment are transformed by somewhat unpredictable processing, and we now have the first tenuous grasp of what these processes may be. We will gradually take our images of the mind and behavior—still set in wooden frames, and composed like the set scenes of Leonardo da Vinci or Rembrandt—and shake them free of the frame. We might begin to imagine memory and behavior as fluid, like the shifting light or the rolling modulations of sound.

Over history we have become clever at using memory, mostly to make money and manipulate social mood. We have experimented with the making of memories on a mass scale in politics and commerce: data is taken, interventions designed, and attitudes moved. There are techniques to steal into our memories through every pore: by television and telephone, by post and radio, by mail and movie and magazine, through the ears and eyes and nostrils. From the false scent of cookies piped into the mall to the instructions for politicians in how to make false gestures, we are shameless in trying to shape what people hold in mind.

What will come as we learn more? The difficulty in living with a new psychology will be some of its choices. If, for example, we find a drug treatment (this is not entirely speculation) that successfully reduces the urge to commit rape in some of those whose urges seem irresistible, what will we do with it? Will prisoners be asked to choose whether to take it? Can such a choice ever be made and not considered coerced? More generally, if a man may make very different choices when he is under treatment than when he is in a "natural," desperate state of mind, which of the choices are we to respect? The natural, free choice to do evil? Or the "treated" choice to do good? To what lengths should we go to transmute the lead of criminality into the gold of social conscience? What will be worth doing? Peculiar things have already resulted from these difficulties, as in the Louisiana case in which a prisoner refused to take his medication for depression. The court had said that, on medication, he could be considered sane, and therefore, he could be executed. Off his medication, he was clearly crazy, and we do not execute the insane, or so the law says.

There are larger issues as well. Richard Lewontin points out in *Biology as Ideology* that as social struggles occur, the weapons of change are speech and demonstration, boycott, strike, and violence, on the one side. And on the other side are the institutions which are "created to forestall violent struggle by convincing people that the society in which they live is just and fair, or if not just and fair, then inevitable, and that it is quite useless to resort to violence. These are the institutions of social legitimization. They are just as much a part of the struggle as the rick-burnings and machinery destruction of the Captain Swing riots in Britain in the nineteenth century. They use ideological weapons. The battleground is inside people's heads; if the battle is won on that ground then the peace and tranquility of society are guaranteed.

"For almost the entire history of European society since the empire of Charlemagne, the chief institution of social legitimation was the Christian Church"; now, as Lewontin says, science is replacing religion as the chief legitimating force in society. It has all the elements of social legitimacy which religion had—it appears to derive from sources outside of human struggle, but must descend as truth from a suprahuman source. Also, its pronouncements seem to go beyond the possibility of compromise or modification. And, as Lewontin slyly says, forces of legitimation must have, as science does, "a certain mystical and veiled quality so that its innermost operation is not completely transparent to everyone. It must have an esoteric language which needs to be explained." Science claims a method that is objective and nonpolitical, true for all time. Scientists, he says, truly believe that except for the "unwanted intrusions of ignorant politicians and the media, science is above the social fray."

(Henry, of course, is in the condition he is because of the legitimacy of science, which lent its authority to the experimenters who worked on him. It is not science that is to blame, for science and religion at bottom are both systems of knowledge and authority. Atrocities are possible under any such system.)

Wherever we go from here, we cannot go back. We move ahead, even to the time when more knowledge leads to more strange social choices, as it did for the prisoner in Louisiana, or for the researcher who hopes to develop drugs to suppress the impulse to rape. I believe there is already a new interest in a variety of cases, in neurology and psychology, philosophy and epistemology, in which defects, or perhaps unusual abilities, throw light upon the landscape of the mind. It has begun with the cases of Alexander Luria, of Oliver Sacks, of Brenda Milner and Suzanne Corkin. A new art and science of the mind is being created as we see the interior landscapes of the conscious and preconscious mind as if for the first time. This terrain is more difficult and strange than we had imagined. In some of these cases, like that of Henry M., we can already see a dim outline of what we will soon know. It is clear we are our mechanisms, but perhaps not as the determinists would have wished.

In thinking about Henry, and the condition into which he has fallen, I find that among my reactions is a faint fear. It is not just the fear engendered by picturing ourselves meeting the same catastrophe that befell him. There is something else. It is a feeling that the incapacity of Henry treads on inviolable ground, like crossing the forbidden earth of the ancestors at night and meeting there something horrible, some feature of ourselves in that dark field. The tale of Henry, after all, is the story of the removal of his humanity. It suggests we are made of removable parts, perhaps predictable units.

There was a time when humans knew little about the inner workings of things human and pictured character as a solid, single entity. Belief and action, thought and memory, all were bound up in a single configuration, a divine reflection. We now find we are not unitary beings, except in our imagination and the stories we tell about ourselves. We are fragmented; we have parts which compete and contradict and cooperate with one another somewhere beneath our attention.

Tales of neurology should now make it clear that we are a bundle of abilities (perhaps in a curious way like the concept of "real estate" which is thought of as property but is actually a bundle of separate rights permitting so much use and no more). Zazetsky, Henry, and Oliver Sacks's various characters all suffered some limited but extraordinary damage to parts of themselves. They are parts which we think of normally as inviolable and part of a unitary "self" which perceives and acts. Neither love nor reason is outside the possibility of being described in mechanical terms. Everything can be atomized. But I think we need not worry. The arguments of dualism may fade as we find the mechanisms beneath behavior, but therein is the lesson:

Alchemists were once convinced that lead could be transmuted to gold, and they devoted extraordinary energy to seeking the secret knowledge that would make it so. Oddly, they were right. In principle, we now know how to change lead into gold. But of course, the point has been lost, because it would cost more to do so than to buy the gold in the first place.

The knowledge the alchemists sought, and which we now have in hand, is more unusual and beautiful than the medieval chemists imagined, but alas they dreamed of riches, and knowledge by itself is no solution to poverty or pain.

Now, with our fiddling in the brain, we have come to a new pass, one that Merlin Donald calls a "neuroscientific apocalypse." As Gerald Edelman describes it, "We are at the beginning of the neuroscientific revolution. At its end we shall know how the mind works, what governs our nature, and how we know the world." Oliver Sacks is also excited, and says of Edelman's theory about the organization of the brain that it is "the first truly global theory of the mind and consciousness, the first biological theory of individuality and autonomy."

The fear as we approach this age of the brain is the fear of reduction, that we will have stripped from ourselves another layer of unknowing and thus lose a layer of protection for our sensitive selves. But the alchemists should remind us that we need not fear; whatever understanding we develop will be far too complex to be a shortcut to perfection or wealth. I recall learning this lesson for the first time some years ago, while working on a different book.

I found myself in the office of a well-known computer programmer at M.I.T., in the artificial intelligence laboratory on Technology Square, interviewing him about a huge computer program he had written, one complex enough to be somewhat unpredictable in operation.

"Can you predict what the program will do in a given situation, when you feed in information?"

"No, of course not, not in any specific way. If I could, there would be no point in building it."

"So you have written a program as big as a library, every line of it from your hand, but you do not know it?"

"Right."

"And cannot predict its behavior?"

"True."

It is thus possible to know all the particulars of a system and still not be able to predict its behavior. Even if we learn the neural code for mankind and can understand the structures underlying thought and feeling, we will not render them predictable. Though somewhat altered, the jobs of psychotherapists will be secure.

This is perhaps the chief lesson of the new biology and what may become the new psychology. Despite the genes, despite the systematic rules under which organisms operate, chance and experience enter at every turn. This makes science difficult, but leaves art free to celebrate the individual experience.

And this is what Oliver Sacks celebrates about the theories of Gerald Edelman, that while they explain much they also demonstrate that each person's experience shapes his or her brain structure so that no two are alike. Because of these differences, each creature with its own history, no model will make humans predictable.

Edelman's theory, whether true in detail or not, has elements which any future theory must have: its base in the wet nets of cells rather than dry computer chips, and its emphasis on each individual adapting to the world, moment to moment. His theory suggests that humans acquired a new kind of memory not long ago in evolution, and this created our reflexive consciousness. The creation of this new kind of memory, in which values are linked to incoming perceptions, creates a conceptual explosion. We can include our own behavior among the items of memory, sketching pictures of ourselves acting in the world, thus standing above all in our own mental worlds, in a way omniscient. The self, the past, and the future are all things which can be pictured, represented, and may interact with one another, as Edelman says; even a consciousness of consciousness is possible. The substrate and servant of this new mentality is memory. And memory is, at bottom, even though challenged by society at times, a possession of individuals.

To the sessions of sweet silent thought,
I summon up remembrance of things past,
I sigh the lack of many a thing I sought,
And with old woes newly wail my dear time's waste.
                          —Shakespeare, "Sonnet 30"

27.

"Do you know who you live with?"

"No."

"Do you know if you live with your parents?"

"No, I'm not sure."

"Do you know where you live now?"

"I don't remember right off . . ."

"Why do you have trouble remembering?"

"Well, what I keep thinking of is that possibly I had an operation. And somehow the memory is gone . . . And I'm trying to figure it out . . . I think of it all the time. I don't remember this, and why I don't remember that."

"Is that worrisome?"

"Well, it isn't worrisome in a way, to me, because I know that if they ever performed an operation on me, they'd learn from it. It would help others."

"Uh huh."

"And possibly they wouldn't make the same mistakes again that they were making with me . . ."

"Do you think about not remembering where you are, or what you're doing? Does that get you upset?"

"It does get me upset . . . But I always say to myself, what is to be is to be. That's the way I always figure it now."

"Do you remember any of the friends that you have now?"

"No. I don't. But I don't worry any. That's the funny part about it, in a way . . ."

"You'd think it would be kind of frustrating, though, not being able to remember where you are living, or who your friends are, or any of those things?"

"I don't think about it that much."

"What about your parents?" [At the time of this conversation I had with Henry, spring of 1992, his father had been in the grave twenty-five years, and his mother fifteen years. He had lived alone in a nursing home for more than twelve years.]

"Oh! That's a big question mark I have right there!"

"Yes? Tell me about it."

"I don't remember fully if we're living in the same house, or have moved and sold that house. It was a house we had down on Crescent Drive . . . with a big empty lot next to us. We fenced it all in so the kids wouldn't come in and play baseball there when we went out . . ."

"But you don't know about your parents now, where you're living, or where they're living, any of that?"

"No. And I have question marks in my mind all the time about them."

"What do you ask yourself?"

"Well, I'm always asking, where are they?"

"Uh huh."

"I don't remember. I try to pull one and one together, but I can't remember where they are. Or if one is dead, and maybe the other is living. But I don't know."

"Uh huh."

"Or maybe they both have gone!"

"Right. Okay. . . . Well, thanks for taking time and talking with me, Henry."

"Well, if it helps, all right."

A little earlier, I was walking down the corridor beside Henry, and Suzanne Corkin was making the usual kind of small talk with Henry.

"Do you know where you are, Henry?"

Henry grinned. "Why, of course. I'm at M.I.T.!"

Dr. Corkin was a bit surprised. "How do you know that?"

Henry laughed. He pointed to a student nearby with a large M.I.T. emblazoned on his sweatshirt. "Got ya that time!" Henry said.

It is of course better that Henry has adjusted himself to the impossible, that he has the grace and wit to absorb it. His acceptance—more, his bemusement at times—is still haunting. But I know, now, how he can be adjusted: the world is as he perceives it. It need not be more, because at any given moment, it is to him whole and entire. We may think him deprived, but as we cannot see infrared, do we feel it missing? The bees can see ultraviolet, which makes flowers utterly different objects to them—do we long to see flowers "as they actually are" to the bees, who can experience them better? The bees, after all, have their own shortcomings. Living things are by nature bound within their skins and within their perceptions, which are only derived pale models of that which is outside them. Memory and fantasy, less exact representations of the world, are less reliable, and may lead to trouble.

At the same time, because they are loosed from the absolutely factual, and can assemble so many different pictures out of the same set of facts, from this comes the release of the human spirit.

Memory, upon which we lean so heavily, is at last not fixed, but moves.

This is what we must conclude. Earlier in the century physics pulled the prop of certainty from beneath us when it certified the insubstantial nature of matter; now neuroscience certifies the insubstantial nature of mind as well. We must depend not on memory, but instead upon judgment, in a physical universe which turns, and in a mental universe which shifts like sand under waves. What we can extract from the experience of these moving tides, and what we can reason on top of that, is all we may rely on.

It was time to take Henry home. That is, to carry him from his bed at M.I.T., where he had stayed a week, to his bed at the nursing home. Henry's life at the surface seems uninteresting. His clothes are generic, picked out for him from some bin, nondescript plaid shirts and plain brown trousers with stretch-band waists, for convenience. He lives in a standard aseptic hospital-like room both at M.I.T. and the nursing home. At the nursing home, he has a roommate, their worlds divided by a white curtain. His bath is bright white tile, shared. Life is unmarked by colors or textures of individuality. Henry is quiet most times, though he smiles readily. His keepers complete the formula with their pleasant and efficient manner. They move from task to patient task, with professional smoothness, expecting that Henry will now need to go to the bathroom and must be reminded, that he will now need to wash, that even though he wonders whether he needs a shave, he has already shaved twice that day. His roommate will now want the television quiz show, and after dinner demented Marie will have to be separated from one or two of the others she chides and pokes. Ancient Rose might make sexual suggestions to Henry, and will have to be warned away. An altogether orderly procession down dim, polished halls toward death.

There is nothing about this place or person to call attention to itself, on an old highway bypassed by the interstate years ago, and not far from the railroad tracks. But beneath this uniform of gray is nonetheless hidden a well of knowledge. It pierces his life, extending down past love, of which he is incapable, to the seat of the mind.

Henry is in some curious way an oracle from whom answers may be sought, though perhaps few of them understood. Like a sibyl his words are often dark until, later, events illumine what was meant. "The Sibyl of Cumae chants fearful, equivocal words," as the *Aeneid* says, "and makes the cave echo with sayings where clear truths and mysteries are inextricably twined. . . They say it is here that the gateway to the underworld is found."

So, we loaded Henry into the car, and got on the interstate to drive him back. As we drove, I tried to think of things to chat about on the road. I asked about the makes of the cars; he saw one he thought must be a new model, a '48 Ford. It was a 1991 Chevy.

I wanted to ask something about what it feels like for Henry. "Can you give me a comparison, what it's like searching around when people ask things?"

"Well, in a way, you're still wondering to yourself, about things, and what has been. You try and think of everything that's going on, or has been. You can't remember that really at all," he said. "It's like waking up, sort of like waking up in the world. You're waking, trying to push things together yourself, reaching back. And you wonder at times yourself just, well, what it is and what it isn't."

"You're aware of what's happening now pretty clearly, aren't you?" I said. "So how far back do you think you can reach? How far back can you remember?"

"Oh, not very far at all."

"So, for example, do you remember getting in the car, anything that happened while when we were getting in the car?"

"I truly don't."

"That was just a few minutes ago."

"A few minutes ago."

"And then, if you look in the other direction, if you look into the future: Do you know when we get where we're going what you'll be doing?"

"No. I keep on thinking that one thing."

"What?"

"That we're going to a clinic and they're going to examine me to find out different things about me that will also help them with other people."

I asked if he remembered the early days, around the time of his surgery. "No," he said, "I try to forget them in a way."

"Why, Henry?"

"You just try to forget them so they won't bother you."

Though we were heading south through the pines, away from Boston and M.I.T., fifteen minutes after leaving Henry was mentally on his way back again. So his life moves in one direction only, and perhaps partly by choice, he is always going to the clinic, always to help other people by what happened to him.

Arriving at the nursing home where he has lived for more than a decade, I asked if he recognized the place.

"No."

"What do you suppose we're doing here?"

"Well, we must be going to visit somebody!"

"Do you know anyone who lives here?"

"Not that I remember."

On entering, the nurses and others who have worked with him and kept him company for years smiled broadly, "Well, hello, Henry!"

Henry smiled back. "Uhh, hello."

There is not a flicker of recognition in Henry's eyes, but there is pleasure. One after another, people say hello as we take him down the hall toward his room. By the time we reach it, he is beaming and shuffling quickly. Though he remembers not a drop, he is pleased.

And so on a spring evening, Henry goes home. We who have spent time plumbing his character and puzzling over his every utterance, musing about what it suggests of how a human may think *sans* memory, as we leave him in his empty white room with his blank squares for mental landscape, we must imagine him forever lost. But nevertheless, we must also imagine him, in some way, not unhappy.

From this man, this diverted life, has arisen science. His life is as a marker, and we, hiking a long dark path, have come up a rise. From it a view opens out as if we had new eyes; we look into a new, varied, and unfamiliar plain where we may walk, refreshed and curious about what lays ahead. Though Henry stays behind, we walk on.

We enter an age in which the hidden structures of brain and behavior—the beams in the walls, the wiring in the ceilings—are laid open to view. I imagine Henry beckoning us toward the door, with his disarming and unaware grin, as he shuffles through himself. Many men and women over many decades have actually brought us to this pass, but Henry's was an extraordinary sacrifice and a crucial one, and I like to imagine him shuffling ahead of us through this next portal of knowledge, then motioning us on as he watches.

Acknowledgments

The most important debt I owe is to those scientists who were willing to spend time explaining their work. Chief on the list of those who aided in this project is Dr. Suzanne Corkin of M.I.T., who spent much time with me, opened her files on H.M., and offered advice generally.

Dr. Brenda Milner was also helpful in describing her experiences with H.M. and helping me to spend time with H.M. himself. Dr. Larry Squire was a continuous fount of information and experience on the subject of memory.

Some of those interviewed, and whose ideas were also important in shaping the book: Dr. Mortimer Mishkin, of the National Institutes of Health; Dr. Endel Tulving, emeritus professor at the University of Toronto; Dr. Daniel Alkon of the National Institutes of Health; Dr. Eric Kandel of Columbia University; Dr. Gary Lynch of the University of California at Irvine; and Dr. Philip Gold of the National Institute of Mental Health.

For her patience, criticism and aid, I owe a debt to my wife, Carisa B. Cunningham.

Bibliography and Notes

I. *General books, or specially interesting books on memory and the mind*

Luria, Alexander R. *The Man With a Shattered World, the history of a brain wound.* Cambridge, Mass.: Harvard University Press, 1987.

———. *The Mind of a Mnemonist, A little book about a vast memory.* New York: Basic Books, Inc., 1968.

Yates, Frances A. *The Art of Memory,* Chicago: University of Chicago Press, 1966.

Alkon, Daniel L. *Memory's Voice.* HarperCollins, 1992.

Zeki, Semir. *A Vision of the Brain.* Oxford: Blackwell Scientific Publications, 1993.

Parkin, Alan J. *Memory.* Oxford: Blackwell Publishers, 1993.

———. *Memory and Amnesia.* Oxford: Basil Blackwell Ltd., 1990.

Rose, Steven. *The Conscious Brain.* New York: Alfred A. Knopf, 1975.

———. *Making of Memory.* New York: Anchor Books, 1993.

Searle, John R. *The Rediscovery of the Mind.* Cambridge: The M.I.T. Press, 1992.

Dennett, Daniel C. *Consciousness Explained.* Boston: Little Brown and Company, 1991.

Penrose, Roger. *Shadows of the Mind.* Oxford: Oxford University Press, 1994.

Restak, Richard M. *The Brain Has a Mind of Its Own.* New York: Harmony Books, 1991.

———. *The Mind.* Toronto: Bantam Books, 1988.

———. *The Modular Brain.* New York: Charles Scribner's Sons, 1994.

Ornstein, Robert. *The Evolution of Consciousness*. New York: Prentice Hall Press, 1991.

———. *The Psychology of Consciousness*. New York: Harcourt Brace Jovanovich, 1977.

——— and Thompson, Richard F. *The Amazing Brain*. Boston: Houghton Mifflin Company, 1984.

Sacks, Oliver W. *A Leg to Stand On.* New York: HarperCollins Publishers, 1990.

———. *Awakenings.* New York: E.P. Dutton, 1983.

———. *The Man Who Mistook His Wife for a Hat*. New York: Summit Books, 1985.

Squire, Larry R. *Memory and Brain*. Oxford: Oxford University Press, 1987.

Thompson, Richard F. *The Brain*. New York: W. H. Freeman and Company, 1985.

Humphrey, Nicholas. *A History of the Mind*. New York: Simon & Schuster, 1992.

Gardner, Howard. *The Mind's New Science*. New York: Basic Books, Inc., 1995.

Corsi, Pietro, ed. *The Enchanted Loom: Chapters in the History of Neuroscience*. New York: Oxford University Press, 1991.

Finger, Stanley. *Origins of Neuroscience: A History of Explorations into Brain Function*. New York: Oxford University Press, 1994.

(The above two volumes contain, in addition to a huge array of drawings, paintings and photographs in the history of neuroscience, historical articles on the development of neuroscience and sketches of the major scientific figures in the field.)

II. Notes on H.M. and scientific papers

The information about Henry in this and following chapters comes from a variety of sources, chief of which is the files of Dr. Suzanne Corkin at M.I.T., where she has collected medical and nursing notes dating since 1946, before the surgery of H.M.; altogether the most complete documentary record of Henry.

The richest but most difficult source is Henry himself, whom I interviewed at length several times. I spent some time with him as he went about his usual routines of testing at M.I.T. and life in his nursing home, and spoke to a number of those who have worked with him extensively, including Neal Cohen, Edith V. Sullivan, and Margaret Keane, all of whom worked at the M.I.T. Clinical Research Center.

I spoke to a colleague of Dr. Scoville, to his daughter, and to his wife of that time. I checked birth and death records in Connecticut and census reports in Louisiana for historical data, and visited the sites in Hartford and East Hartford he mentions, a few of them with Henry himself there to narrate.

A listing of some of the most interesting scientific papers about Henry follows:

1954. Scoville, William Beecher, "The Limbic Lobe in Man." *The Journal of Neurosurgery*, vol. 11, p. 64-66. (This is the first reference to H.M., a few passing words only, received by the journal on October 12, 1953, six weeks after H.M.'s surgery.)

1955. Penfield, Wilder and Milner, Brenda. "Memory Deficit Produced by Bilateral Lesions in the Hippcampal Zone." *American Medical Association Archives of Neurology and Psychiatry*, May 1958, vol. 79, pp 475–497, (Read at the 80th Annual Meeting of the American Neurological Association, June 13, 1955, in Chicago, but not published until 1958.)

1957. Scoville, William Beecher and Milner, Brenda. "Loss of Recent Memory after Bilateral Hippocampal Lesions." *The Journal of Neurology, Neurosurgery and Psychiatry*, 1957, vol. 20, p. 11. (The first substantial paper on H.M. himself.)

1959. Milner, Brenda. "The Memory Defect in Bilateral Hippocampal Lesions." *Psychiatric Research Reports* 11, American Psychiatric Association, December, 1959. (Discussion follows article.)

1965. Corkin, Suzanne. "Tactually-guided maze-learning in man: effects of unilateral cortical incisions and bilateral hippocampal lesions." (Suzanne Corkin's first paper on attributes of H.M.)

1966. Milner, Brenda. "Amnesia Following Operation on the Temporal Lobes." In *Amnesia*, C.W.M. Whitty and O.L. Zangwill, eds. Montreal Neurological Institute, reprint number 870., published by Butterworths, London, 1966. (Here Dr. Milner notes that the first time the middle temporal parts of the brain were found to be critical to memory was in a paper presented in 1899 by V.M. Bekhterev, and published in 1900, "Demonstration Eines Gehirns mit Zerstorung der vorderen und innerin Theile der Hirnrinde beider Schlafenlappen," in the *Neuro Zentbl.*, vol. 19, p. 990-1. Milner says, "At a medical meeting in St. Petersburg, he demonstrated the brain of patient who had shown a severe memory impairment as her earliest and most striking clinical abnormality; the main pathological finding was bilateral softening in the region of the uncus, hippocampus and adjoining mesial temporal cortex." She notes other, similar reports later in the literature, beginning in 1947 and 1952.)

1968. Scoville, William Beecher. "Amnesia after Bilateral Mesial Temporal-lobe Excision: Introduction to Case H.M." *Neuropsychologia*, 1968, vol. 6, pp 211–213. (This same issue of *Neuropsyhologia* contained an introductory paper by Milner beginning on p. 175, four articles on studies of H.M., and other articles on memory after brain lesions. The article on p. 215 was by Milner, Suzanne Corkin, and Hans-Lukas Teuber, the three people who spent more of their time with H.M. than any others. The book collected material from a September, 1966, conference held in Lagonissi, Greece, by the International Neuropsychological Symposium.)

1984. Corkin, Suzanne. "Lasting Consequences of Bilateral Medial Temporal Lobectomy: Clinical Course and Experimental Findings in H.M." *Seminars in Neurology*, vol. 4, number 2, p 249–259. (This is a concise summary at the time of the work that had been done on H.M.)

1991. Squire, Larry R. and Zola-Morgan, Stuart. "The Medial Temporal Lobe Memory System." *Science*, Sept. 20, 1991, vol. 253, p. 1380–1386. (This is a landmark paper and a relatively recent, thorough summary of knowledge about what the hippocampus and its related parts in the middle temporal area of the brain do in memory.)

III. Bibliography and notes by chapter

CHAPTER 1.

Shakespeare, William. *The Complete Works*. Stanley Wells and Gary Taylor, eds. Oxford: Clarendon Press, 1992.

Proust, Marcel. *In Search of Lost Time*. Translation revised by D.J. Enright. New York: The Modern Library, 1992.

Johnson, Samuel. *Selected Poetry and Prose*. Frank Brady and W.K. Wimsatt, eds. University of California Press, 1977.

———. *Selected Writings*. Patrick Cruttwell, ed. Middlesex: Penguin Books, Ltd., 1968.

Bate, W. Jackson. *Samuel Johnson*. New York: Harcourt Brace Jovanovich, 1977.

Dryden, John. *John Dryden*. Keith Walker, ed. Oxford: Oxford University Press, 1987.

De La Mare, Walter. *Collected Poems*. London: Faber and Faber, 1990.

James, William. *The Principles of Psychology*. George Miller, ed. Cambridge: Harvard University Press, 1983. See especially chap xvi., "Memory."

———. *The Writings of William James*. John J. McDermott, ed. Chicago: University of Chicago Press, 1977.

Snyder, Solomon A. *Drugs and the Brain*. New York: Scientific American Library, 1986.

CHAPTER 2.

See notes above on Scoville, Milner and Corkin.

Eliade, Mircea. *Myth and Reality*. Willard K. Trask, tr. New York: Harper & Row, 1975.

Anonymous. *Sweeney Astray*. Seamus Heaney, tr. New York: Farrar Strauss Giroux, 1983.

Bruner, Jerome. *In Search of Mind*. New York: Harper & Row, 1983.

CHAPTER 3.

Alkon, Daniel L., MD. *Memory Traces in the Brain*. Cambridge: Cambridge University Press, 1987.

———. *Memory's Voice*. *supra*.

Thompson, Richard F. *The Brain. supra.*

Humphrey, Nicholas. A *History of the Mind. supra.*

Kandel, Eric R. and Schwartz, James H. *Principles of Neuroscience.* New York: Elsevier, 1985.

Allport, Susan. *Explorers of the Black Fox.* New York: W. W. Norton & Company, 1986.

CHAPTER 4.

Schank, Roger C. *The Connoisseur's Guide to the Mind.* New York: Summit Books, Simon & Schuster, 1992.

————. *Dynamic Memory.* Cambridge: Cambridge University Press, 1990.

———— and Abelson, Robert P. *Scripts, Plans, Goals and Understanding.* Hillsdale, New Jersey: Laurence Erlbaum Associates, Publishers, 1977.

Jaynes, Julian. *The Origin of Consciousness in the Breakdown of the Bicameral Mind.* Boston: Houghton Mifflin Company, 1990.

Anon. *The Epic of Gilgamesh.* N. K. Sandars, tr., ed. Middlesex: Penguin Books, Ltd., 1976.

CHAPTER 5

Schmandt-Besserat, Denise. *Before Writing, vol. I and II.* Austin: University of Texas Press, 1992.

Ong, Walter J. *Orality and Literacy.* London: Routledge, 1990. On oral history, including issues of memory.

Ascher, Marcia and Robert. *Code of the Quipu.* Ann Arbor: The University of Michigan Press, 1981.

Le Goff, Jacques. *History and Memory.* New York: Columbia University Press, 1992.

Plato. *The Dialogues of Plato, vol. I and II.* Benjamin Jowett, tr. New York: Random House, 1937. See Theatetus, vol. II, p. 195; Phaedrus, vol. I, p 278-9; Phaedo, vol I, p. 456 ff.; Meno, vol. I, p. 360 ff.

CHAPTER 6.

The Bible. King James Version. Oxford: Oxford University Press, 1972.

Confucious. *The Wisdom of Confucious.* Lin Yutang, ed. New York: The Modern Library, 1938.

Aristotle. *Introduction to Aristotle.* Richard McKeon, ed. New York: The Modern Library, 1992.

Lucretius. *The Nature of Things (De Rerum Natura).* Frank O. Copley, tr. New York: W. W. Norton & Company, 1977.

Cicero. *On the Good Life.* Michael Grant, tr. London: Penguin Books, 1971.

Snodgrass, Joan Gay, Levy-Berger, Gail, and Haydon, Martin. *Human Experimental Psychology.* New York: Oxford University Press, 1985.

CHAPTER 7.

Augustine, Aurelius. *The Confessions of Saint Augustine.* John K. Ryan, tr.

New York: Doubleday, 1960.

Yates, Frances A. *The Art of Memory. supra.*

CHAPTER 8.

Priest, Stephen. *Theories of the Mind.* Boston: Houghton Mifflin Company, 1991.

Corsi, Pietro, ed. *The Enchanted Loom: Chapters in the History of Neuro-science. supra.*

Finger, Stanley. *Origins of Neuroscience: A History of Explorations into Brain Function. supra.*

CHAPTER 9.

Ribot, Theobule. *Diseases of Memory in Significant Contributions to the History of Psychology, 1750–1920.* Daniel N. Robinson, ed. Washington, DC: University Publications of America, 1977.

Ramon y Cajal, Santiago. *Recollections of My Life.* E. Horne Craigie, tr. Cambridge: The M.I.T. Press, 1991.

James, William. *Writings. supra.*

Ebbinghaus, Hermann. *Memory.* Henry A. Ruger and Clara E. Bussenius, tr. New York: Dover Publications, 1964.

CHAPTER 10.

Bergson, Henri. *Matter and Memory.* N. M. Paul and W. S. Palmer, tr. New York: Zone Books, 1988.

CHAPTER 11.

Valenstein, Elliot S. *Great and Desperate Cures.* New York: Basic Books Inc., 1986.

Hippocrates. *Hippocratic Writings.* G. E. R. Lloyd, ed. Middlesex: Penguin Books, Ltd., 1983.

CHAPTERS 12-14.

See notes in 2. above for Scoville, Milner and Corkin.

CHAPTER 15.

Luria, Alexander R. *Man with a Shattered World. supra.*

———. *Mind of a Mnemonist. supra.*

———. *The Working Brain.* New York: Basic Books, 1973.

Tulving, Endel. Transcripts of interviews with Kent Cochran, from 1984 to 1986, kindly supplied by Dr. Tulving.

———. *Elements of Episodic Memory.* Oxford: Clarendon Press, 1985.

CHAPTERS 16-19.

Heaney, Seamus. *Poems, 1965–1975.* New York: Noonday Press, Farrar, Straus and Giroux, 1980.

———. *Selected Poems, 1966–1987.* New York: Noonday Press, Farrar, Straus and Giroux, 1990.
These two volumes contain many poems whose figure or substance is memory.

CHAPTER 20.
Nabokov, Vladimir. *Speak, Memory.* New York: G. P. Putnam's Sons, 1968.
Thomas, Elizabeth Marshall. *The Hidden Life of Dogs.* New York: Houghton Mifflin Company, 1993.
There is growing a small literature, written by patients or their doctors, reflecting on various human abilities and the loss of them. Of course, Dr. Oliver Sacks is the chief expositor of these subjects in his several books mentioned above. Here are others I have found:
Cytowic, Richard E. *The Man Who Tasted Shapes.* New York: G. P. Putnam's Sons, 1993.
Hull, John M. *Touching the Rock: An Experience of Blindness.* New York: Vintage Books, Random House, 1992.
Rymer, Russ. *Genie: An Abused Child's Flight from Silence.* New York: HarperCollins Publishers, 1993.
Williams, Donna. *Nobody Nowhere: The Extraordinary Autobiography of an Autistic.* New York: Times Books, 1992.
Kaysen, Susanna. *Girl, Interrupted.* New York: Turtle Bay Books, Random House, 1993.
Grealy, Lucy. *Autobiography of a Face.* Boston: Houghton Mifflin Company, 1994.
Doernberg, Myrna. *Stolen Mind: The Slow Disappearance of Ray Doernberg.* Chapel Hill, N.C.: Algonquin Books, 1986.
Schaller, Susan. *A Man Without Words: A True Story of Psychological Breakthrough.* New York: Summit Books, Simon and Schuster, 1991.
Renee. *Autobiography of a Sciozphrenic Girl: Reality Lost and Regained.* Grace Rubin-Rabson, tr. and Marguertie Sechehaye, providing an analytic interpretation. New York: Signet, New American Library, 1970.
Rapoport, Judith L. *The Boy Who Couldn't Stop Washing: The Experience and Treatment of Obsessive-Compulsive Disorder.* New York: Plume, New American Library, 1990.
Sudnow, David. *Ways of the Hand: The Organization of Improvised Conduct.* New York: Bantam Books, 1979.

CHAPTER 21.
Squire, Larry R. *Memory and Brain. supra.*
Squire, Larry R. and Nelson Butters. *Neuropsychology of Memory.* New York: The Guilford Press, 1984.

CHAPTER 22.
Donald, Merlin. *Origins of the Modern Mind.* Cambridge: Harvard University Press, 1991.

CHAPTERS 23–24.

Neisser, Ulrich. *Cognition and Reality.* New York: W. H. Freeman and Company, 1976.

———. *Memory Observed.* New York: W. H. Freeman and Company, 1982.

CHAPTER 25.

Loftus, Elizabeth. *Memory: Surprising New Insights Into How We Remember and Why We Forget.* Reading, Massachusetts: Addison-Wesley Publishing Company Inc., 1985.

——— and Ketcham, Katherine. *Witness for the Defense.* New York: St. Martin's Press, 1991.

Wright, Lawrence. *Remembering Satan.* New York: Alfred A. Knopf, 1994.

Crews, Frederick. *Victims of Repressed Memory.* New York: The New York Review of Books, Parts, I and II, plus correspondence, beginning December 1, 1994.

CHAPTER 26.

Lewontin, R. C. *Biology as Ideology.* New York: HarperCollins Publishers, 1993.

Edelman, Gerald M. *Bright Air, Brilliant Fire.* New York: Basic Books, 1992.

———. *The Remembered Present.* New York: Basic Books, Inc., 1989.

Index